BIOLOGY, ZOOLOGY, AND GENETICS

Evolution Model vs. Creation Model

Adell Thompson
University of Missouri—Kansas City

UNIVERSITY
PRESS OF
AMERICA

University Press of America, Inc.™
P.O. Box 19101, Washington, D.C. 20036

Printed in the United States of America

ISBN (Perfect): 0-8191-2922-4
ISBN (Cloth): 0-8191-2921-6

TABLE OF CONTENTS

LISTS OF ILLUSTRATIONS AND CHARTS

Preface

Biology, Zoology, and Genetics: Creation Model vs. Evolution Model is a textbook intended to give students an insight into the controversy as it now stands. Teachers, administrators, school boards, and parents are all being asked to take sides on this issue.

Creationists also have adjusted their arguments to another abiding judicial fact. The U.S. Supreme Court found in 1968 that laws enforcing the teaching of the creationist view were in effect laws that involved religion and therefore were unconstitutional. The First Amendment provides that "Congress shall make no law respecting an establishment of religion or prohibiting the true exercise thereof..."

To get around that barrier, creationists have tried to sever the religious element from creation doctrine as they propose it for the schools: the theory is that the universe and everything in it were created, but it doesn't say by whom.

In a statewide election three years after the Scopes trial in Tennessee, Arkansas adopted a law making it a crime to teach evolution in public schools.

The Tennessee law of Scopes' day wasn't scrapped until 1967, a few months before the old Arkansas law fell in the U.S. Supreme Court, First Amendment ruling favoring Little Rock biology teacher, Susan Epperson. The court said: "The over-riding fact is that Arkansas' law selects from the body of knowledge a particular segment which it proscribes for the sole reason that it is deemed to conflict with a particular religious doctrine; that is, with a particular interpretation of the Book of Genesis by a particular religious group."

In the 56 years since Scopes' and the 13 since Epperson, creationist science memamorphosed, creationist legal theories evolved.

Now the disciples speak of "only an appearance" of age in the universe, explain geological formations by the cataclysm of Noah's day, see change "within kinds" but nothing like evolution.

In their attacks on evolutionists, creationists like to cite Mark Twain's ironical shaft at evolution:

"There is a fascinating thing about science. One gets such wholesale returns of conjectures out of such a trifling investment of facts."

The major purpose of this book is a dual one. First, to focus upon the two models in the major areas of life sciences--Biology, Zoology and Genetics; and second, to provide students in the sciences with a realistic look at the creationist view (as it compares with the evolutionary theory). I have been a teacher for 24 years and as most of my counterparts am satisfied with using the evolutionary theory as a major theme throughout my courses. To my knowledge, with the exception of a few writers espousing one model or the other, there has been no real effort on the part of scientists or textbook writers to suggest giving both models equal consideration in a textbook. I am therefore presenting the facts as they have been interpreted in my research of the literature.

I wish to personally thank Ms. Linda Werner, a graduate student who assisted me in the literature search, particularly giving the views of the creationists with biblical scriptures they have used to make their case justifiable. Many of the views suggested by evolutionists are common knowledge to most biologists, thus they have been limited in scope unless it was necessary to contrast the two models.

CHAPTER I

INTRODUCTION

CREATION AND EVOLUTION MODELS DEFINED

There are two doctrines regarding biological change--the doctrine of evolution and the doctrine of special creation. Those subscribing to evolution, which is the amoeba-to-man thesis, adopt the theory that all living forms in the world have arisen from a single source that came from an inorganic beginning. So, according to the model of evolution, the first living cell "evolved" into complex multicellular forms of life; these "evolved" into animals with backbones. Fish evolved into amphibia, amphibia into reptiles, reptiles into birds and mammals, early mammals into primates, and primates into man. Thus, there was a gradual change fo simple forms into more and more complex forms.

Subscribers to special creation adopt the theory that the basic kinds of plants and animals were placed here on earth by the direct action of the Creator; and say that, during time, created forms, or kinds, have changed in size or quality; yet the changes have been within certain boundaries of the original kinds (still not clearly discerned) which were made by the Creator.

It is a fact that neither Creation nor Evolution can be observed today or reconducted in the laboratory, and we cannot go back in time to see how the origin of life occurred. Creation and Evolution, are not as testable as the laws of Thermodynamics, yet they are both said to be theories. A theory results from testing a hypothesis for which an abundance of evidence has been accumulated to form a major conceptual scheme. It must be capable of generating hypotheses which are not only testable but also falsifiable. Evolutionists contend that evolution fulfills these conditions on every count. Perhaps the term "theory" may not be adequate in describing creation and/or evolution. The terms "models" or "doctrines" could perhaps describe them more accurately, because in order to accept the theory or doctrine of either creation or evolution, knowledge is required--knowledge of what perhaps did occur in past history.

Though we cannot show perfect proof, we can examine the evidence and records in the fossil record--a key to the past--and see which model is most logical given this evidence.

If we observed the complexity of life even in its simplest form, from the DNA molecule found in all cells, we see that the DNA alone is so complex that we wonder how chance or random action of matter could produce such complexity. There is great complexity and precision in the tiny cells as well as in the highly complex systems found in man. Darwin himself said, when he thought of the eye and its development from organisms that had no eyes or a peacock's tail and its development, it made him sick. He could see that it would be difficult for eyes (for example) to develop from structures that had no eyes.

Despite the certain complexity of things, we can find order in them. In the law of cause and effect (i.e., for every action there is an equal and opposite reaction) we see the necessity for a first cause. There had to be a cause--a beginning. Creationists suggest that the Bible concurs with the first cause provision in claiming that there was a sudden appearance in a great variety of highly complex forms.

> "In the beginning, God created heaven and earth." (Gen 1:1) The tremendous complex of orderly relationships in the universe, challenging the highest intelligence of man to describe, certainly implies an intelligent First Cause of these relationships.[1] The evolutionists still contend there was a gradual appearance of simple life forms developing to the more complex. The fossil records confirm this belief.

One long-range reason I had for writing this book was to become, thereby, equipped to teach a more accurate combination of the facts of science. When studying science in all fields--biology, zoology, genetics, geology, astronomy, etc.--we should use all knowledge available to us and should present all of this to our students. Students should be exposed to both sides of the coin regarding biological change--the doctrine of creation and that of evolution--since both are based on models. We should examine the evidence in light of all knowledge we have--scientific knowledge (facts), philosophic knowledge (opinions), religious knowledge (revealed knowledge)--and then allow students to draw their own conclusions.

I realize that discussion of origins is not, strictly speaking, science, because origins are not subject to experimental verification. There were no scientific observers present when life began, and these events are not taking place now; so solutions to the problem of origins are impossible by scientific means alone.

BODIES OF KNOWLEDGE

We have three major areas, or bodies, of knowledge. One is scientific; the second is philosophic; the third, religious.

Science is only one kind of knowledge that we possess. Some people like to imply that all knowledge is scientific. But there are other realms of knowledge. Scientific knowledge is knowledge acquired with our senses. Scientific knowledge is the kind we can feel, see, and touch. It is observable. For example, we can put water in a pan on a stove and observe at what temperature it boils.

Philosophical knowledge is a kind of knowledge that is a matter of opinion. For example, we have knowledge of the importance of several occupations to society--doctor, lawyer, fireman, etc. But to say which is more important cannot be "proven" scientifically; so it becomes a matter of opinion. This is beyond the scientific realm of knowledge.

Then there is the knowledge we could not have unless it were revealed to us from a supernatural source. This is religious knowledge.

Is it possible, scientifically, to know the origin of life: with our human senses and abilities to come to a scientific proof of the origin of life? Some say yes, others say no!

From a philosophical standpoint, we see the Egyptians' thought life originated in the Nile River, the Greeks' from the mud, and the modern scientist believes that billions of years ago in an ancient primeval soup some hot amino acids were formed. The molecules came together and formed DNA and this was the beginning of life.

> Well, the third area of knowledge, religion, tells us that "In the beginning God created heaven and earth" (Gen. 1:1) and that on the third, fifth, and sixth days...various forms of life...were created. On the sixth day, God formed man of the dust of the earth and breathed into him "the breath of life and man became a living soul." (Gen. 2:7) This knowledge we could not have unless it had been revealed to us.[2]

You may not scientifically prove God created man, or that life originated in the ocean or mud. Yet scientists continue to probe the origin question. We must utilize all areas of knowledge (scientific, philosophic, and religious).

In the book Historical Geology Carl O. Dunbar writes:

> From the earliest times, men have speculated upon the origin of the world in which we live and legends of a creation are known to almost every race and tribe. Many of these beliefs are fantastic, but their very existence shows that the earth's beginning has been an endless challenge to the minds of man, that every race and people have legends of a creation.[3]

SCIENTIFIC METHOD

Next, I would like to examine the scientific method in brief. The following are steps in the scientific method. When the scientist is especially aware of certain initial observations in one area of study and perplexed about certain aspects (1) he states a problem, or asks a question that no one else has studied or solved. Then (2) he gathers many facts that may have a bearing on the problem. Next (3) he forms an hypothesis, or estimate, that may help explain the problem. (4) More facts are gathered, and their relevence to the hypothesis is weighed. Then, if possible, (5) experiments are performed. If the gathered facts are consistent with the suggested explanation or hypothesis, the scientist comes to the conclusion that his explanation is valid, and (6) he publishes his results.

If the suggested explanation becomes established as a result of efforts by many research workers who repeat the steps and arrive at the same conclusions and rule out other explanations, then the explanation becomes or is called a theory. When proved without any shadow of doubt, it may be called a scientific law, principle, or doctrine.

A theory is a very well established hypothesis (tentative solution to a scientific problem, which evidence may confirm or disprove) that has been thoroughly tested. Neither creation nor evolution can follow the procedures in the scientific method directly. Origins of life can neither be repeated or tested in the laboratory; so both are assumptions. However, in most schools--elementary, high school, and college--evolution is introduced as an established theory and the alternate doctrine of creation is not mentioned and is regarded as unscientific.

Scientists place limitations on their work and deal only with observations and study of things that can be demonstrated by and with the senses. The scientist doesn't deal with God as a subject of science but with laws and principles God established. He does not deal with miracles, but only with events occurring with principles he's uncovered.

He, therefore, does not deny God's existence or miracles, but many times excludes them from his research work.

FIRST AND SECOND LAWS OF THERMODYNAMICS

There are two laws of science accepted by all scientists: The Law of Conservation of Energy, that says that matter or energy can neither be created nor destroyed but can be changed from one form to another: that energy remains constant. This is the First Law of Thermodynamics.

The Second Law of Thermodynamics is the concept that while energy cannot be destroyed, it becomes useless and unavailable to do work. All ordered systems left to themselves become disordered, and the general tendency is downward. Another way to describe this is that all processes manifest a tendency toward decay and disintegration which is lost; thus this indicated a net increase in what is called entropy, or state of randomness or disorder, of the system.

These two laws which most scientists accept as the most universally applicable principles of science were only recognized about one hundred years ago. Many question how the man-made elements may affect these laws.

These two laws, though expressed in scientific terminology, have also been expressed in the Bible for thousands of years, according to the creationists. The conservation principle is clearly set forth by the fact of a completed creation which is now being sustained by the Creator.

> Colossians 1:16, 17 illustrates both aspects of this truth, "By Him were all things created...and by Him all things consist." The word "created" is used in the past tense, so it, creation, is not presently (sic) going on. The word consist comes from a Greek word from which we get our English word sustain. So the verse is saying, in effect, that "By Him all things are sustained." "By the Lord Jesus Christ all things--systems, structures, all kinds of organisms and relationships were created once and for all in the past and now being conserved."[4]

The first chapter of Genesis describes the Creation, and it is here, in the Bible, where we can obtain information about creation methods and any other details about it. The evolutionists contend that knowledge revealed by the Bible is untestable, thus has no scientific foundation. Scientific study of present processes will not be able to reveal anything about creation except that it took place. Evolutionists assumed that these present processes are those by which all things have developed from primeval chaos into their present complexity. They also suggest that creationists say nothing about open or closed systems of retrogressive evolution.

As well as having the conservation principle, we also have the decay principle. The Second Law of Thermodynamics is like the First, a universal law governing all processes. Even though energy is never destroyed, it continually becomes less available for further work.

Everything seems to eventually wear down and/or die out, to run down, disintegrate, and/or die. All processes do involve change, but the change is in an upward direction, the evolutionists suggest.

There is no question that the Second Law is universal in both the physical and biological worlds as far as science can determine. Living things left to themselves collapse, deteriorate, grow old, and eventually die. This is the destiny of the universe: that all energy of the sun and stars will be degraded and will be unable to be converted into useful work. Even though evolution generates an approach to its phenomena, it is a consequence of open thermodynamic systems. It generates questions about the details of the histories of species, the mechanisms of natural selection, and the relations among natural systems. By contrast, creationism does not possess these qualities.

Creationists feel there is a universal process of change going on in the world, but the real law of change is decay and not growth--a change down and not up, or as an evolutionist once said, "Molecules over eons of time change into men who sustain life for a short period."

According to Huxley, a leading author of the evolutionary world, evolution is an irreversible process moving upward and onward into more and higher levels of complexity and order.

Donald E. Chittick speaks of the Second Law of Thermodynamics. He says, that, given two events, since the whole universe is running down, that event that has most disorder is one that comes second. So, we can see the direction of time by watching the flow of disorder. Disorder always increases with time.[5]

Science contends that energy can be neither created nor destroyed, but can only be changed from one form to another; hence the total amount in a system remains the same, regardless of the physical or chemical changes it may undergo. The Second Law of Thermodynamics says that in any energy transformation, dissipation tends to occur. Thus, after each energy transformation, the available energy in a system will be slightly lower than before. As a result, according to this law, all processes act so as to "increase entropy." Finally, according to both models, this entropy is a measure of disorder; this means that the system becomes increasingly disorganized as the available energy decreases.

IMPACT OF EVOLUTION UPON SOCIETY

Evolution, in this discussion, entails the concept that the complex life we see today, including man, came from a lesser complex form many ages ago. Even further back in time, the lesser complex forms had developed from single-celled organisms.

> "The idea of evolution has been adopted and has been intruded into or has intervened in the thinking patterns of all the main desciplines of man's knowledge. For example, the following disciplines can be included: economics, history, philosophy, sociology, jurisprudence, literature, as well as science."

> Thomas Barnes further says, "The feeling that evolution is the answer, is so widespread that the interpretive framework of history, sociology, economics, and even science itself has been adapted to accomodate evolution.[6]

If one majors in social studies or the humanities, he is being selectively indoctrinated into adoption of the evolutionary point of view.

In history, some believed in evolution so much that they felt they could see evolutionary application to the political life. This thinking was patterned after the thinking of Karl Marx. He adopted evolution as a "so called" scientific socialism. But Charles A. Beard, who wrote a book (circa 1913) called An Economic Interpretation of the United States Constitution, can specifically be referred to here. Beard used this interpretation to express what he thought he would find from researching property ownership records --the voting records of men and women living in the thirteen colonies: that the U.S. Constitution was approved as a consequence of class warfare. (However, he admitted he did not do the actual research.) He maintained that class warfare was involved when the rich and landed gentry and the titled individuals of the colonies supported the Constitution; whereas the poor farmers, debtors, and criminals did not.[7]

The importance of this is that "class warfare," as used by Karl Marx, is a direct outworking of the "so called" struggle for existence, which many people who supported Darwin's ideas utilized to explain the natural selection that he, himself, proposed as the means for the occurrence of evolution. So, in this nation from 1913 to the 1940's, people learned the Beard point of view--an interpretation of the history of the United States in keeping with Marxian ideas based on evolutionary premises. This meant that the history of America was taught to people who became teachers, and whose students, also became teachers.

In literature, evolution in turn, also has had an impact. One example is the author, Jack London, who is an honest supporter of socialism. He used his novels to make attractive to his readers the concepts and ideas of socialism. In The Call of the Wild, he applied the concept of the struggle for existence in evolution which is similar to the concept of class warfare that Marx used in Scientific Socialism.

George Bernard Shaw's dramatic works which have had great impact on the intelligentsia of England and the U.S.A., present evolutionary thinking and the socialistic point of view.

Creationists suggests it is strange that our "so called" beautiful letters are in themselves purveyors and a means for selective indoctrination of the young of many successive generations into an evolutionary point of view. Shaw was a member of the Fabian Society in England.

Another example of the impact of evolution is found in the organizations of governments and societies. As mentioned, Karl Marx was a man who adopted and absorbed the evolutionary point of view and the natural selection and struggle for existence theories found in Darwin's thoughts and writings. It is believed by the creationists that the organization of communist and socialist dictatorships are twins of the same philosophy, and are rooted in the adoption of evolutionary thinking through the writings of Karl Marx.

THE CONNECTION BETWEEN EVOLUTION, MARXISM, AND COMMUNISM

Evolution indicates that there has been a transformation from one animal kind to another animal kind or from one plant kind to another plant kind. Men have remarked that organisms have evolved from simple, to two, to many cell layers, and fish have become amphibians, or frog-like. Frogs have become snakes, and snakes have become birds and human beings. The 1959 edition of Darwin's Biological Work, quotes the evolutionist's contention that:

> On an unprejudiced view we may say the evidence partly supports evolution and partly separate and independent creation."... "Is there any positive proof from any part of the evidence that evolution has or has not occurred? There is no visible proof nor any kind of certain proof either way anywhere.[8]

What is the evidence that we have to confirm evolution? The answer is that we have evidence for variation within limits.

What is Marxism? Marxism is the so-called Scientific Socialism that Marx wrote about and developed as a series of publications in The Communist Manifesto, Das Kapital, and other works.

According to J. Edgar Hoover, "Socialism is the ownership by the state of the means of production and distributions...In general, Socialism would mean an end to private ownership and control of wealth." How does this relate to evolution? A journal that is favorable to Socialistic ideas suggested that Darwin's Origin of the Species was an exciting book to Marx and Engles. In fact, when Marx first read Origin of the

Species in 1860, he wrote to his friend as follows: "This is the book which contains the basis and natural history of our view." And a few weeks later, he wrote to another friend, "Darwin's book is very important and serves me as a basis in natural science for the class struggle in history." We can thus see a possible connection among evolution, socialism, and Marxism.

Marxists profess to seek the same goal whether they be Socialistic, in the sense of Marx, or whether they be dialectic materialists, in the sense of the Lenninist Communist frame of reference. They may differ in their means of approaching the same goal, but in final analysis they acknowledge that there is no individual citizen and that the Marxist doctrine leaves no room for a concept of inalienable rights given to man from birth by his creator, which any serious citizen in the United States recognizes is a key part of our American founding documents. It is important to point out that Marxian biology is today practiced in Russia. Politicians, not scientists, seem to hold the upper hand in biology--which further explains the prominent role evolution plays in biological thinking in the Soviet Union today.

THE IMPACT OF EVOLUTION ON SOCIETY

We can see that evolutionary philosophy is having an impact in most fields of higher learning and major walks of life. Truly in psychology, the theory of evolution has had an effect, implying that the fittest people survive. In psychological testing, for example, the Distribution Theory for the bell curve is based on distribution of mutations which Darwin discovered in the theory of survival of the fittest and is supportive of his theory of evolution.

This evolutionary philosophy is found in political levels and is slipping into our religion. Sects and cults are beginning to be prevalent in our country with the philosophies of Karma, which maintains that if you are perfect in this life you will come again (reincarnation) as more intelligent or in a higher position in life the next time around. Again, the fittest and the best survive. Humanism and many other "isms" have evolutionary philosophy as their base.

Evolution has infiltrated all fields of education, society, and even religious thought.

After having researched this subject, I see the theory of evolution as man's attempt to find his own truth. It reminds me of what men believed before Columbus proved the world round. The word of God should be a challenge and a hope for the scientist to find answers that are so obscure. The word of God says that all were created after their own kind. There is evidence of sudden appearances of various kinds. Thus, the creationist's point of view is supported by the Bible. The fossil record supports the evolutionist's point of view.

FOOTNOTES

[1]Henry M. Morris, The Remarkable Birth of Planet Earth (California: Institute for Creation Research, 1973), p. 2.

[2]H. Douglas Dean, Creation Counseling--Life in a Testube, Vol. 4, Tape 2. (California: Creation Life Publishers, 1973).

[3]Ibid.

[4]Morris, op. cit., p. 14.

[5]Donald E. Chittick, Creation Counseling--The Age of the Earth, Vol. 3, Tape 4. (California: Creation Life Publishers, 1973).

[6]Thomas Barnes, And God Created--Origins Without God and The Beginning of Life, Vol. 3, Tape 2. (California: Creation Life Publishers, 1973).

[7]John N. Moore, And God Created--Impact of Evolution on Society and Evolution and Communism, Vol. 2, Tape 3. (California: Creation Life Publishers, 1973).

[8]Ibid.

CHAPTER II

BIOLOGY

DEFINITION

Biology is the branch of knowledge that deals with the study of living things. What is life?

Although there is to be found no complete definition of the remarkable entity called life, it is, however, helpful to consider the properties of living things and ways we can learn about them.

The following are ten characteristics of living things: (1) living things have definite size and shape; (2) living things are chemically active; (3) living things have a life span; (4) living things originate, grow, age, and die; (5) living things are organized in units called cells; (6) living things require a source of energy; (7) living cells are composed of protoplasm; (8) living things are able to reproduce; (9) living things respond to stimuli; (10) living things have a critical relationship to environment.

CHEMICAL BASIS OF LIFE

As we look around us, we see matter: walls, trees, soil, water, and air are all matter. Matter at times is in motion: a stream flows, the wind blows, animals run, and plant life reaches toward the light. The question might be asked: Can all things in the world be classified as matter or motion? Democritus, a Greek philosopher, believed the following: He thought that all action, including that of animals and even man, is the result of the laws of matter: that action is not brought about by purpose and planning, but only by the result of action that has gone before.

Aristotle, also a Greek philosopher, did not deny the existence of matter and motion, but believed there was some kind of extra something in living matter that activated it.

In the Middle Ages, some people believed that substances in living things were of a different nature from the stuff in non-living matter; that these substances could be studied in laboratories by the same methods used on metals, acids, bases, etc. They did not have empirical data as a basis for this belief, but held it subjectively.

In the 19th century, Frederick Wöhler did an experiment that proved this idea. In 1828 he heated ammonium cyanate and got a substance called urea. It was just like the waste substance found in the urine of animals. The process Wöhler used, however, was a simple one; while the formation of urea in an animal's body is one of an extremely complex series of changes. The ammonium cyanate Wöhler used was made of substances produced by animals. At a later time, a student of Wöhler's formed ascetic acid from carbon, chlorine, sulfur, and water. In this case, an organic substance that was previously believed to be made by living things alone was produced in a lab from inorganic materials.

Therefore, living things do owe their distinctive qualities to the elements they are composed of, because they are made of and contain the same elements as non-living matter. The exact organization of these elements in living matter is hard to determine for as soon as the analysis begins, the tissue dies and great changes take place. A chemist can discover the compounds in non-living matter, but he cannot always know all of the organization of the living things.

Antoine Lavoisier (1773-1794) contributed much to the science of chemistry in his studies of combustion and the atmosphere. He discovered that burning phosphorus gained as much weight as was lost by the air in the container surrounding it. He found, too, that when one fifth of the volume of air was consumed, no more burning occurred even though there was unburned phosphorus present.

Joseph Priestly (1733-1804) also made studies of gases and discovered that when a candle was burning in a closed container, it soon went out. The same result was obtained when a mouse was placed in a closed container of air. It died soon after it was covered. Priestly further experimented by heating the compound mercuric oxide and produced the gas we know now as oxygen. Priestley's contemporaries did not understand the chemical change that was involved; even Priestly himself did not grasp the full importance of his discovery. He saw that the mouse lived longer in this new gas than it would have lived in the ordinary air. He found that the new gas did not harm animals; and when he breathed it himself, he found it refreshing. When a candle was in this gas, it burned rapidly with a bright and noisy flame.

Lavoisier began to make more tests of his own after learning of Priestly's experiments. He concluded that Priestly's new gas, oxygen, composes 1/5 of our atmosphere and that nearly 4/5 of it is the gas we now know as nitrogen. He also showed that the oxygen in the air unites with other substances when they burn.

There were some scientists who were convinced that this process, now known as cellular respiration, was completely chemical and that life itself could be explained as a purely chemical process. The nature of life might be investigated, then, by chemical means. They began to believe that life itself was nothing more than matter and motion.

Further studies brought some doubts about this conclusion's finality. A superficial likeness does not always reveal an essential likeness. Although both combustion and cellular respiration require oxygen and produce carbon dioxide, the two processes are very different; and the burning of a candle is not always parallel to the metabolism of a mouse.

Lazarro Spallanzani (1729-1799) did an experiment that some believed showed life to be no more than chemistry. He was aware that foods become liquid in the stomach, and knew also about the acidic gastric juice he found in bird's stomachs. He put some very small pieces of mutton in a test tube and wheat grains in another and added the gastric juice taken from turkeys. He then sealed the tubes. In order to keep the contents warm as they would have been in a turkey's stomach, he put the tubes under his arms. After three days had passed, the meat and wheat were partially digested and there was no sign of decay. He added more gastric juice, which completely digested the food materials. He then as a control repeated the experiment using water instead and the meat and grain spoiled and did not digest.

Since the chemical process of digestion took place in a test tube away from a living body, some scientists concluded that this was evidence that all life processes are purely chemical. If we give attention to the production of enzymes, we can see that their formation is affected easily by emotion. For example, a grieved or angry person may be so affected that he is unable to eat. He will not have an appetite and his food will disagree with him. He may even vomit. These are examples of biochemical reactions to situations outside the body. When a child is sick and in the hospital, the gastric juice in his parents' stomachs may be decreased. It is true that gastric juice contains a chemical whose action can occur outside the stomach, but the influence of an emotion upon its production is more than a simple reaction. No chemical action outside living cells is affected by emotional reaction, according to some scientists.

Another remarkable fact should be noted: Gastric juice easily digests even tough bits of meat. However, it does not digest the stomach wall which is also made of protein and is not as tough as the meat it digests. When a person dies, the gastric gland stops secreting gastric juice and that gastric juice already present digests the wall of the stomach. So, the living stomach resists self digestion while the dead stomach does not. This suggests that life may be more than chemistry or chemical reaction. These reactions are the agents of life.

In the activities of living things we see qualities not found anywhere else. The growth of a tree differs from the addition of outer layers on a stalagtite or an icicle. The tree takes in carbon dioxide and salts and changes them into protoplasm--bark and wood; yet neither the icicle nor the stalagtite make any change in the materials added to them.

In the body of a person or an animal matter comes and goes, but the individual form remains and continues. The atoms come in as food and leave as waste. It has been estimated that the body is renewed every seven years. Though the true time necessary for this renewal cannot be measured, we know the body is not the same matter it was at one time. Furthermore, genes--the hereditary determiners--are no exception. The body is renewed according to the same pattern, and thus people you know recognize you. We might ask what continues? It has to be something over and above matter: a pattern, not atoms.

It would be difficult for chemistry to account for all the activities of animals. Some of their actions seem to be the result of their own desires, and it is difficult to classify them as coming about because of stimulation by outside forces. Animals, when caused to react, have a choice of reactions. The simple amoeba even displays this choice. When touched by a needle, it may retreat as fast as possible, or roll up into a ball. If held down with a needle, the amoeba just pinches off this portion of flesh and gets away.

Reproduction goes beyond physics and chemistry. Nothing smaller or more simple than a living cell can do this. A virus, which some people say is alive and others say is not, cannot reproduce itself outside a living cell, but depends on a host to do so.

Animals and plants are composed of matter and are influenced by motions. Also in their being seems to be something more. Though life is difficult to define, it seems certain that it transcends matter.

CELLS

When you peel an onion, and pull the sections apart, you can see the structure. Each section of the onion has tiny units called cells. The yolk of an egg is a large single cell before the young starts to develop. These units are exceptions, since most cells cannot be seen without a microscope.

Even as early as the 17th century, men were using a specially prepared glass called a lens to see very small things. The glass was ground to be thicker in the center than at the edges, giving it a convex shape. An object held close to the lens appeared to be larger (magnified).

Anton Van Leewenhoek learned to grind lenses to give high magnification. He viewed drops of water from a river and saw things no one had seen before.

Englishman Robert Hook arranged several lenses together in a metal tube and thus made a compound microscope that gives more magnification than just one lens. He viewed many things--minerals, textiles, and small living creatures.

Hook looked at a thin slice of cork and found it was made of little box-like compartments. They made him think of little rooms in a large building like a monastery. He called these little rooms "cells", as they are called in monasteries. What he was really seeing were the non-living walls of the plant units of which all the living parts had died. It was a century and a half before scientists learned the importance of cells and that they exist in all living things.

R. J. H. Dutrochet (1776-1847) published a paper in 1824 saying that all living things are composed of cells. All early studies by Hook and Dutrochet were done with plants; therefore, at that time, there was no evidence that animals were made of cells.

When viewing the cells of orchids, Robert Brown (1773-1858) noticed a dark sphere that he termed the "nucleus". Other men in science noted around the nucleus the thin, jelly-like substance we call today the "cytoplasm". Together the nucleus and cytoplasm make up protoplasm--the living matter in the cell. Noted also was a thin membrane that bound the cytoplasm. Other observers noted that in plant cells the cytoplasm usually surrounds a central cavity or vacuole containing a fluid cell sap.

M. J. Schleiden, a German botanist, in 1838 came to the conclusion that all plants are comprised of cells. He talked of his hypothesis with Theodore Schwann, a zoologist, who in 1839 made the suggestion that animals also consist of cells; together they developed the cell principle, or theory, that all living things are composed of cells. This was in accord with the idea of Dutrochet fourteen years earlier. The processes of photosynthesis (synthesis by means of sun light; the process by which plants manufacture food) and cellular respiration are examples of biological processes involving energy functions. These are typical cell functions; thus the concept emerges that all living things are composed of cells.

Animal cells have shapes which vary according to functions they perform. For example, striated muscle cells are long and slender. Nerve cells have spherical shapes with thin fibers that, in humans, can be more than two feet long.

The outer layer of a cell can be a selectively permeable membrane and allow only certain substances to pass through. Sometimes there are small vacuoles which contain various non-living materials. The centriole is visible near the nucleus and is a small body involved in cell division. Mitochondria release energy from food and are cell factories. Golgi bodies, shaped like plates, are found within the cell. Other than their influence in secretion, the function of golgi has not been fully determined. Ribosomes, so tiny that they are only seen in electron microscope pictures, are involved in forming proteins. They are attached to an extensive system of tubes known as the "endoplasmic reticulum". The function of the tubes is not completely understood, but some scientists believe they conduct the materials through the cells in an orderly way and provide a working surface for the action of enzymes.

The nucleus is larger than other cell parts and has a double membrane of its own to surround it. In it is a dense body, the nucleoulus (little nucleus). The rest of the nucleus is made of chromosomes and a fluid matrix.

Chromosomes have a shape like flexible rods or pieces of cord. It is difficult to see them when cells are not dividing. They occur in pairs usually, and the members of all chromosome pairs, except the sex pair, are of equal length. In each body cell of a human being are twenty-three pairs of chromosomes. The number differs in various animals. Chromosomes contain the hereditary factors--the "genes". They are composed of proteins and nucleic acids.

In plant cells the nucleus, Golgi bodies, and mitochondria are similar to those found in the cells of animals. However, only plant cells have cell walls which are composed of cellulose. Inside the wall is a thin cell membrane like that found in the animal cells. There is no visible centriole in the plant cell, but cell division takes place in like manner as in the animal cell. (Figure 1)

A much prominent feature in mature plant cells is the large central vacuole filled with cell sap. The

ANIMAL CELL
(Figure 1)

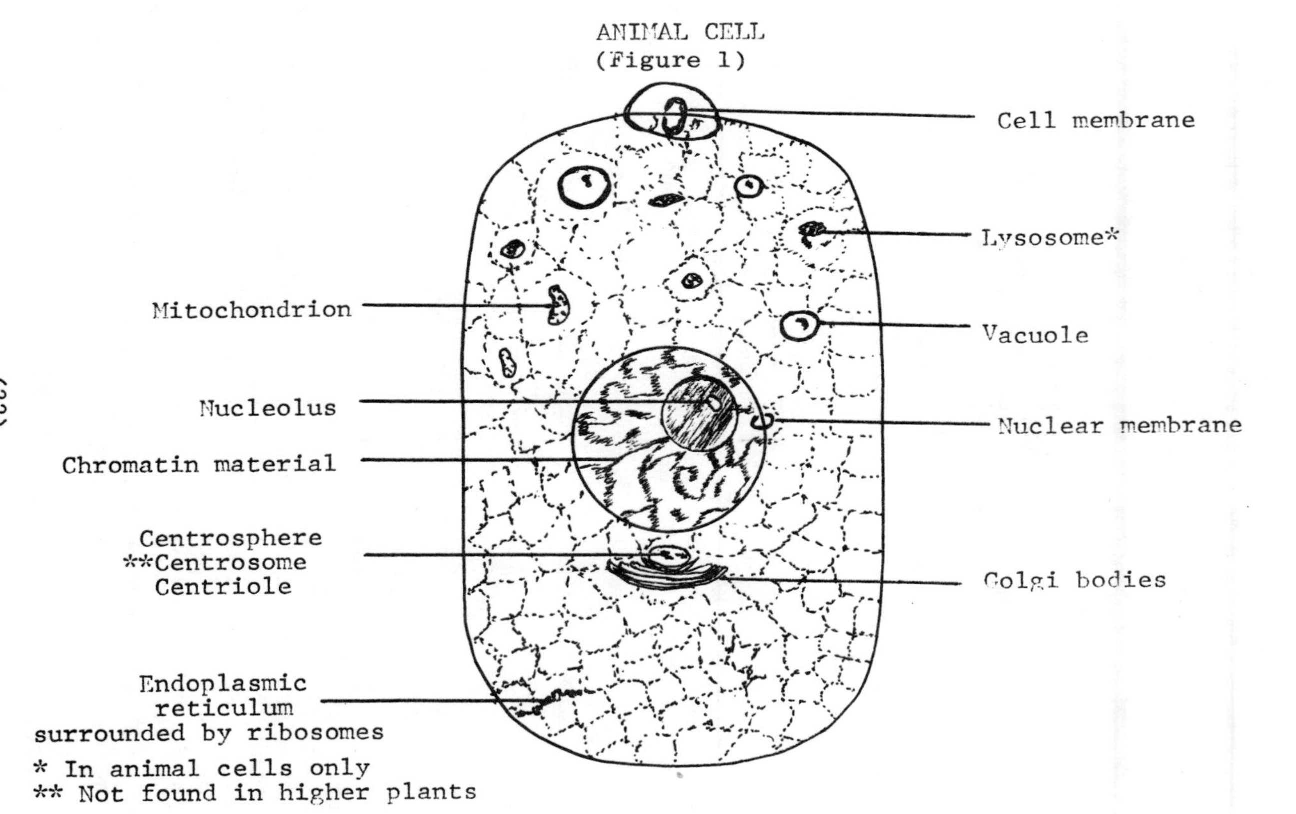

(Figure 2)

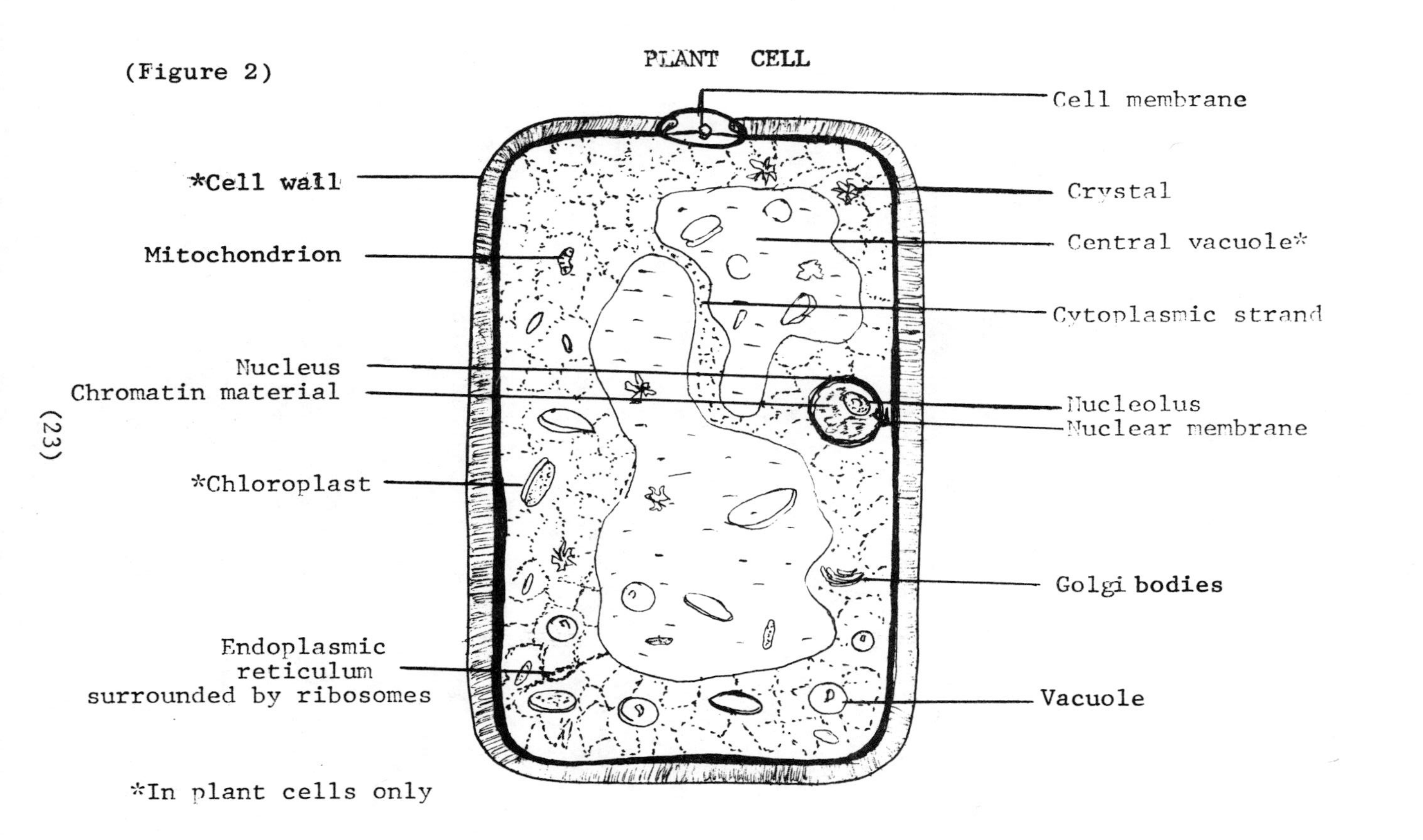
PLANT CELL
Cell membrane
*Cell wall
Crystal
Central vacuole*
Mitochondrion
Cytoplasmic strand
Nucleus
Chromatin material
Nucleolus
Nuclear membrane
*Chloroplast
Golgi bodies
Endoplasmic reticulum surrounded by ribosomes
Vacuole
*In plant cells only

protoplasm surrounding the vacuole can easily be mistaken for part of the cell wall. (Figure 2)

In green plants, chlorophyl pigments are found in dense bodies known as "chloroplasts". The chloroplasts function in making glucose (a sugar) with sunlight as the source of energy. This is photosynthesis. The glucose and other foods made from it by the cell are used by plants and ultimately by almost every living thing.

Schwann believed--in regard to the origin of cells--that, when an egg developed into an embryo, its yolk simply divided into units of proper size and formed into nuclei and other cell parts.

Several scientists believed cells originated in the division of other cells, and in no other way. As an organism grows, each of its cells divide and form two new cells. Sexual reproduction occurs when the sperm of the male unites with an egg from the female, forming a zygote from which an embryo grows. In this instance, as in others, life is shown to come from previous life.

Mitosis is the regular method of cell division in plants and animals. The chromosomes contain many genes which govern the various cell traits. Mitosis is a detailed and orderly process in which each new cell comes to possess the same kinds and numbers of chromosomes present in the parent cell. This process exemplifies order in complexity. (Figure 3)

Chromosomes become easier to see in the nucleus when cell division begins. When they become visible, they are stretched out long and slender. They may have already split into two duplicates and the daughter chromosomes stay joined together.

The centriole divides into two new centrioles, and these begin to move to opposite ends of the nucleus, while the nuclear membrane and nucleolus begin disintegrating (prophase).

Chromosomes become shorter and thicker and line up in a single plane midway between two centrioles (metaphase). If viewing them from the side, they appear to be broad straight lines. Lines or rays appear all around the centrioles like light rays from

ANIMAL MITOSIS
(Figure 3)

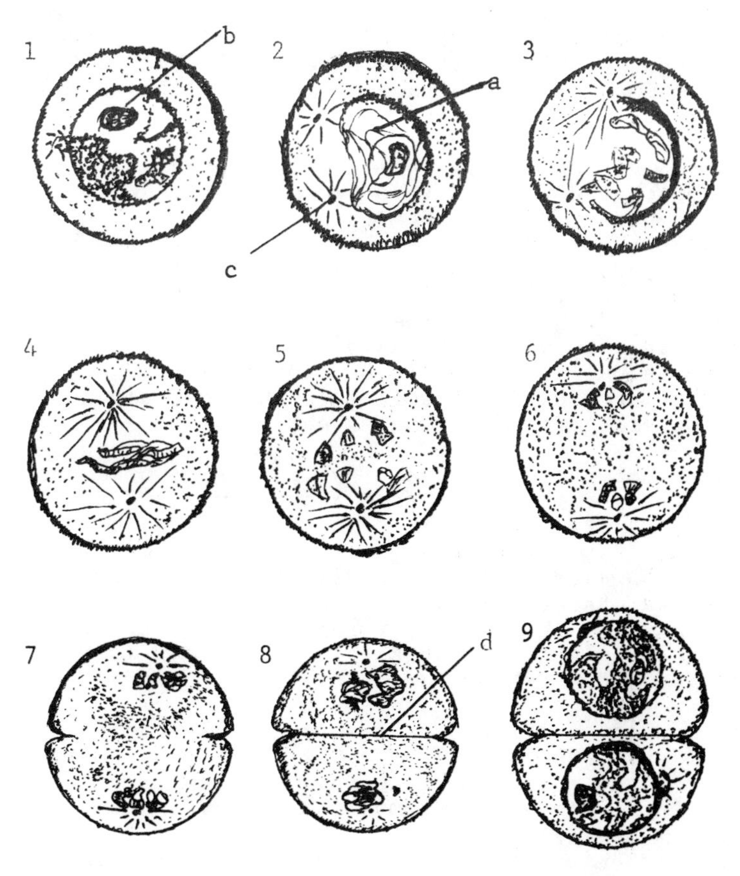

1. Interphase
2. Early prophase
3. Late prophase
4. metaphase
5. Early anaphase
6. Late anaphase
7. Early telophase
8. Late telophase
9. Daughter cells

a. Centromere
b. Nucleolus
c. Centriole
d. Cleavage furrow

a star. The rays extend toward the chromosomes and attach to the chromosomes. These are called spindle fibers. It is possible that the spindle fibers contact and pull the daughter chromosomes, one of each pair going toward each of the centriole pairs, some to the right side and others to the left, forming two equal clumps of chromosomes and becoming the centers of two daughter cells. If the parent cell had forty-six chromosomes (in man), each daughter cell likewise would have that same number.

This division is accomplished very precisely. The chromosomes duplicate into two daughter chromosomes; and tests indicate that the genes form exact copies of themselves, making the genes in one daughter chromosome just like those in the other daughter chromosome. There is a definite spot on each chromosome where the centriole fiber attaches. Also, if one daughter chromosome is pulled to the right, the other one is pulled to the left. The above steps insure that each daughter cell gets the same amount of genetic material found in the parent cell.

After the chromosomes are pulled into two clumps, nuclear membranes begin to form around them and nucleoli are seen to form again. The rest of the cell is pinched into two parts that are equal (cleavage furrow). In plants, however, there is a difference. A straight partition separates the daughter cells, and there is no pinching from the outside inward. The centriole is absent in the plant cell, but the fibers lead to points at opposite sides of the parent cell (cell plate).

As mentioned before, each species has a definite number of chromosomes...If both the sperm and the egg contributed the full number to the young animal, there would be too many chromosomes and genes, but the problem has been solved in a remarkable way. As the eggs are being formed in the ovary or the sperm being formed in the testes, there is a pair of cells dividing which produces cells with only half the number of chromosomes. This process is an exceptional pair of divisions known as meiosis. When the sperm and egg join in fertilization, the normal number of chromosomes is restored. Man has twenty-three pairs of chromosomes, a total of forty-six.

Forty-six is the diploid number for man, even though the human sperm has only twenty-three chromosomes, the monoploid number. Meiosis is the formation of gametes (reproductive cells), a process of chromosome reduction resulting in cells with one set of chromosomes. (Figure 4)

We can understand how meiosis occurs. For example, we can think of a cell in an animal's reproductive organ having the normal diploid number of chromosomes at the beginning of its division, the chromosomes then divide logitudinally. As in meiosis, the daughter chromosomes still lie beside each other. As the spindle fibers begin to form, something unusual occurs: each double chromosome moves alongside another double chromosome structually identical to it.

One chromosome comes from the father and the other from the mother. Recalling that the chromosomes have split, there are four strands now lying side by side. This is called synapsis. The four strands twist and have been observed to break and exchange portions during synapsis.

The entire cell begins to divide into two. The synapsed chromosomes separate and go into the newly-forming nuclei as in mitosis. But the daughter chromosomes do not pull apart; they remain double. Another cell division takes place later in which there is no splitting of chromosomes, but the daughter chromosomes separate from each other. The cell formed is thus monoploid and has one chromosome of each pair.

Meticulous provision has been made for the continuance of cells and of the kinds of plants and animals. One is impressed with these complex processes. It seems unreasonable to think that such processes occur by chance or accident.

BIOGENESIS AND SPONTANEOUS GENERATION

There are certain processes that only living organisms can perform, one of which is the unique process of reproduction. Plants reproduce plants, animals reproduce animals, and human beings are the descendents of human beings. This process of life yielding life is known as biogenesis.

People throughout all time have asked the question: Is reproduction the only way an organism can come into being? Scientists have gathered much data,

(Figure 4) ANIMAL MEIOSIS

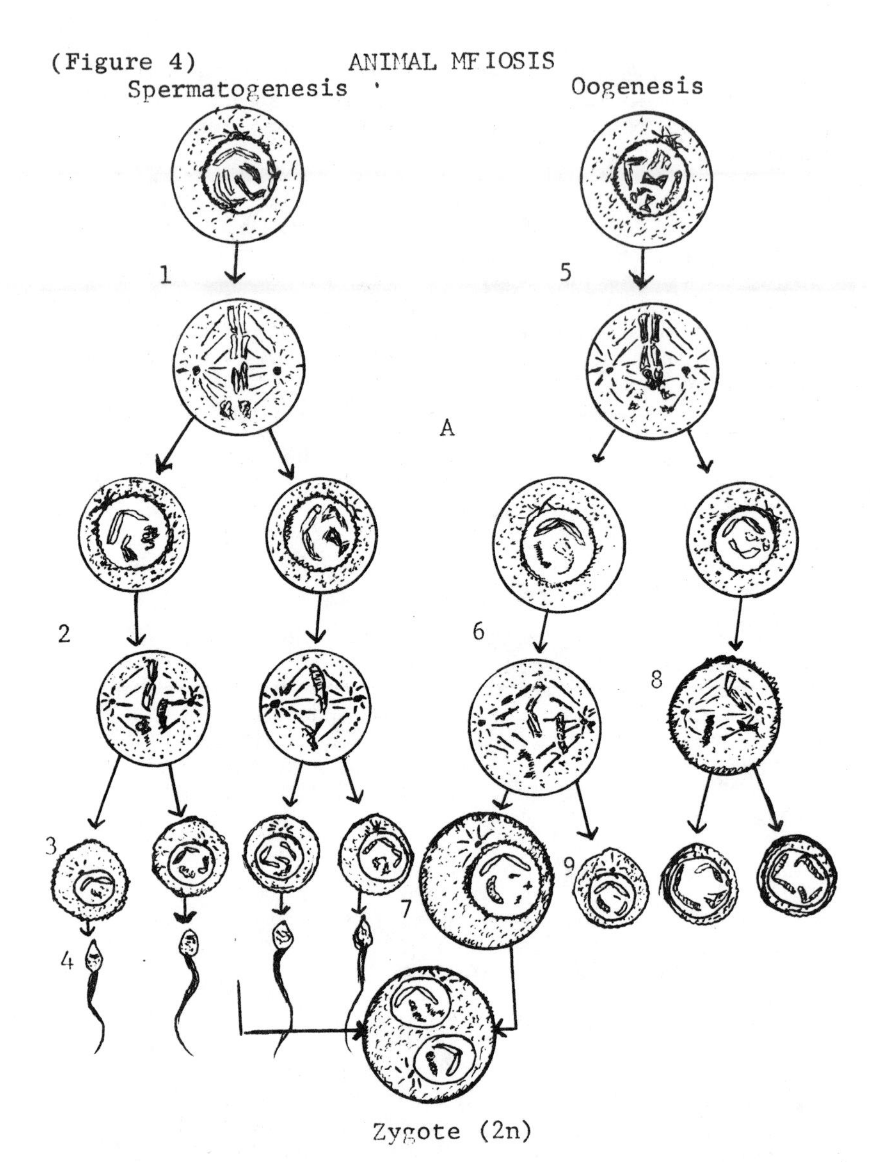

A. First meiotic division
B. Second meiotic division

1. Spermatocyte I (2n)
2. Spermatocyte II (2n)
3. Spermatids (n)
4. Spermatozoa
5. Oocyte I (2n)
6. Oocyte II (2n)
7. Ovum (n)
8. Polar body 1
9. Polar body 2

and if one relies on currently known processes, there is no doubt that life does come into being by reproduction from life. Historically, the viewpoint that life comes from life has been so well established through facts discovered by experimentation, it has come to be known as the Law of Biogenesis. There are no alternative means of generation of life in so far as present scientific laws are concerned.

In ancient times, people widely held the belief in spontaneous generation. When the Nile River would flood its delta, numerous mice plagued the country of Egypt. It was believed and claimed that they arose directly from mud. The ancient Egyptians were not aware that meadow mice had two to six litters a year and that each litter consisted of six to thirteen young.

Even Aristotle believed that many animals, for example frogs, and crabs and worms, came directly from soil. Aristotle was a brilliant philosopher who came to many accurate and valid conclusions, but people did not know enough to correct his mistakes by further observation and experimentation.

A man named J. B. Helmont believed that putting wheat, cheese, and soiled linen together in a jar would spontaneously produce mice. The idea of life coming from non-living matter is known as spontaneous generation.

Louis Pasteur helped rid biology of the idea of spontaneous generation. Some believed meat exposed to the air would produce flies, and pond water, frogs,etc.

Spontaneous generation became unpopular because of scientists like Francesco Redi. Redi said meat will not produce flies unless adult flies, parent flies, lay eggs upon the meat. Others disproved that mice arise from wheat, or frogs from pond water, but that they require parents to produce the young, and that there is an entire life cycle. Redi proved this idea with his simple experiments that allowed flies to lay eggs in containers, and he watched them develop into maturity.

But while the famous Louis Pasteur was alive, many scientists still believed that microscopic bacteria could arise spontaneously. They believed that these tiny creatures could come out of the food that was present for them in the bottles in which they grew. So the spontaneous generation theory wasn't really dead.

Pasteur did not believe in spontaneous generation, and he wanted to disprove it by experimentation. To test the theory he began by sterilizing bacterial food or broth inside a flask. After sterilizing it, he sealed it and no bacteria formed in the bottle. The food remained clear, indicating no bacteria were present. Pasteur had opposition from others who suggested that he rigged the experiment. They believed that spontaneous generation would have occurred if the life-giving powers of air had been present. So Pasteur, stimulated by new ideas, decided to try to do an experiment having sterile bacterial food in a bottle open to the air, and yet not allow dust particles to get into the liquid and contaminate it with bacteria. Pasteur believed it was the dust that carried the germs into bottles. He constructed a flask with a narrow neck at the top that went off into a curve at the side, something like an "S" shape These flasks were named after him and are known as Pasteur flasks because they were open to the air, allowing air to enter the bottle but, the flasks did not allow dust particles to enter the bottles as they were trapped in the bend of the glass. He then repeated this experiment, sterilized the food for the bacteria, and kept it in the special flasks. He found that the food inside the bottle remained crystal clear, showing that no bacteria had formed. These same bottles are on display at the Pasteur Institute in France.

Pasteur wanted to disprove the spontaneous generation theory because of what he believed about God. He said,

> "This problem of spontaneous generation is all important and all absorbing. It is the very problem of life and its origin. To bring about spontaneous generation would be to create a germ. It would mean to go from matter to life through conditions of environment and matter. God, as author, would then no longer be needed. Matter would replace Him. God would need to be involved as only author of the motions of the world and universe."[9]

He then said, and therefore, gentlemen, "I could point to that liquid and say to you, I have taken my drop of water from the immensity of creation, and I have taken it full of the elements appropriated to the development of inferior beings. And I wait, and watch, and question it, begging it to recommence for me the beautiful spectacle of the first creation. But it was dumb, dumb, since these experiments were done several years ago; it is dumb because I have kept it from air, from life, for life is a germ and a germ is life. Never will the doctrine of spontaneous generation recover from the mortal blow of this simple experiment."[10]

It is true that spontaneous generation is now rejected by biologists. Evolution focuses upon this disproved theory of spontaneous generation, when discussing the law of biogenesis, one of our basic principles. People once believed that life came from non-life long ago in the early ocean.

The law of biogenesis does not involve the origin of life in the beginning. How did life come about in the past? Science cannot answer this question because the origin is beyond the reach of science. Some may say there was no beginning and that somewhere in the universe life has always been present. But the universe must have had a beginning for the energy of sun and stars is being given off continuously, and if there was no beginning, they would have lost their heat or mass (source of their heat) long ago in eternal existence.

ORIGINS OF LIFE

Special Creation Origins

The creationists have attempted to base their evidences of the origin of life around Biblical quotations, such as: "In the beginning God created heaven and the earth." (Gen. 1:1). This important scriptural statement sets the pace for the entire revelation to follow. In this verse, God answers questions that have for centuries burned in the hearts and minds of men. Who or what is responsible for this world? This first verse of the Bible claims to lay the basis for answering these questions.

In John 1:3, the Bible tells us that He who was "the Word made flesh" (through the virgin birth), "who dwelt among us" was the actual creator of the universe. It says, "All things were made by him; and without him was not anything made that was made." Paul confirms this when he says,

> "For by him were all things created, that are in heaven, and that are in earth, visible and invisible, whether they be thrones or dominions or principalities, or powers; all things were created by him, and for him." (Colossians 1:16)
> It is further confirmed that he created (Hebrews 1:2) when it said, "Hath in these last days spoken unto us by his Son, whom he hath appointed heir of all things (we who believe in him are joint heirs with him) by whom he also made the worlds."[11]

A most important aspect of the supernaturalism of the original creation was its suddenness and instantaneousness. The evolutionary concept of a gradual buildup of heavier and heavier elements throughout cosmic history (for example, in Gamow's "Big Bang Theory") is clearly excluded by scripture, according to the creationists.

They suggest that the Hebrew word "bala" means the sudden bringing into existence supernaturally of highly complex entities apart from the use of pre-existing materials. It is a concept that cannot be experienced in our world because Genesis 2: 1-3 clarifies that God in six days finished His work, ended it and "rested on the seventh day from all His work He had made." Creation is not now going on, so it cannot be studied as a scientific discipline. We depend on His word. It is by faith, faith in this record, and the authority of the God who gave it that we understand that the worlds were framed by the word of God, so that things which are seen were not made of things which do appear (Hebrews 11:3). The visible universe of mass energy does not owe its origin to a previous simpler form of matter energy, but came into being suddenly out of nothing by the spoken word of our personal God. Acceptance of this position rests on an individual's faith in the infinite power of Jesus' use of supernaturalism as the method to bring the heavens and earth into existence.

Whitcomb believed, in the past, that God's word and modern concepts of science were combined to bring about creation. He thought God used processes through long periods of time to bring fruit trees into existence as he does today. God could create seeds from which trees would grow. But God commanded earth to bring forth not seeds but fruit trees yielding fruit after his kind with its seed itself. He created full grown trees, suddenly, heavy with fruit with the seeds in the fruit. He did create seeds, yet they were in the fruits hanging from full grown trees. This is a spectacular aspect of the doctrine of creation--the methods that God the creator used in bringing this fantastic world of living things into existence.[12]

For example, creationists suggests that the human race could have come gradually over hundreds of millions of years of gradual change and have developed from animal kingdoms to this present form. This would need to be a series of miracles to counteract the Second Law of Thermodynamics, which is the only observable process we see in the present world. On the other hand, God may have created Adam as a baby to grow up, or perhaps have created a mother; but where would she come from?...The obvious answers to this problem of the logical point for the commencement of the life cycle of living organisms is the adult. God created Adam and Eve suddenly, as adults, by supernatural methods not observable in our present world. They were capable of bringing forth children and of caring for them just as today.

So, if we were answering the question--which came first the chicken or the egg--God obviously created the hen first, capable of laying and hatching her own eggs. And thus the cycle of life began with adult, full-grown fruit trees; adult animals, and Adam and Eve as adults.

These concepts carry over into other parts of the creation, as when God created the stars as lights and the coal deposits, etc. He created them with an appearance of age, and man then would have use of them.

Creationists suggest that the idea of sudden appearance thus completely saturates the entire creation account, and there is really no alternative to this from a logical standpoint when the First and Second Laws of Thermodynamics are taken into account.

They contend that when God created living things, he chose to use previously created organic substances for the bodies of these creatures. The earth brought forth vegetation, and the living creatures and man came from the dust of the earth, and he commanded the waters to bring forth marine creatures and birds on the fifth day. But the water, of itself, contributed nothing to the complex physical structure and life principle of these animals, just as the water Jesus used at Cana of Galilee could never have turned into wine, even if it vibrated with evolutionary anticipation in those stone jars for millions of years. In the above instances, complex entities appeared all of a sudden even though built upon pre-existent lifeless materials. The supernaturalism and suddenness of creation provide a necessary background for the concept of creation with a superficial appearance of history or age.

The doctrine of the creation of things with an appearance of history is seen throughout the scriptures. However, this doctrine of scripture has met with much misrepresentation and ridicule by many--more so than most other doctrines.

Henry Morris believes that anything created by God, even the simplest of atoms or other creations, would have an appearance of some age. No genuine creation of any type could be without an initial appearance of age inherent in it. And if God could create atomic stuff with an appearance of age--in other words, if God exists--then there is no reason why He could not, in full conformity with His character of truth, create a whole universe full grown.

For example, in all of Jesus' miracles the same methods are seen in action. If the doctrine of creation with appearance of history is erroneous, then most of the recorded miracles of the Jesus could not have occurred. On the side of a mountain near the Sea of Galilee five thousand men and their families ate loaves of bread and fishes that were created. There were tens of thousands of barley loaves composed of grain that neither had been harvested from fields nor baked in ovens. And there were at least ten thousand fishes that had never hatched from eggs or been caught in nets or dried in the sun.

The creation doctrine is clearly seen when Jesus began his public ministry on earth and performed his first miracle, which was intended to manifest his glory (John 2:11) as the world's creator (John 1:3, 14) He did this by transforming about one hundred fifty gallons of water into delicious wine. Wine is the end product of a long series of complex natural processes which involves the taking or drawing of the water from the soil into the fruit of the grapevine. This water is gradually transformed into the juice of grapes. After this, the ripened grapes need to be picked, the juice squeezed out, and the sediments allowed to settle. But Jesus, the Lord of Creation, by-passed the natural and human processes and created the end product with an appearance of history.

> Now it is imperative to note, says Whitcomb, "that the ruler of the feast, who knew not whence it was coming naturally assumed that this good wine had been somewhere kept...until now."[13]

This was a natural conclusion, of course, for neither he nor anyone else in the world had ever considered the possibility of wine coming directly from water. It must, therefore, have had a history of natural development. This is one of the underlying reasons for all denials of supernatural creation.

As people contemplate the created works of Jesus, whether they be heavenly bodies, earth, sea, plants, animals, people, the natural man, as the ruler of the feast, assumed that they have all been "kept" somewhere "until now," having gone through complex natural processes, from simple forms over long periods of time.

Other examples of miracles point to this doctrine of appearance of history. In John, chapter nine, John tells of a man with congenital blindness who was healed and could see perfectly. The rulers of Israel did not believe he was blind until they asked the man's parents. The healed man himself said, "Since the world began was it not heard that any man opened the eyes of one that was born blind." (John 9:32) In a moment Jesus created the appearance of a man born with normal eyesight.

In a like manner, Lazarus was raised and had the appearance of a man who had not died, yet had been in the tomb four days earlier.

We could come to the conclusion that, if God created living things after their kind, as the first chapter of Genesis reports ten different times, He could have created them with a superficial appearance of age. The Bible says to us that God began the cycle of life with fully mature adult organisms, not embryonic forms who then produced after their own kind.

Dr. Thomas H. Leith has attacked this position in a paper entitled, "Some Logical Problems with the Thesis of Apparent Age," that he presented at the Nineteenth Annual Convention of the American Scientific Affiliation in August, 1964. He claimed the doctrine lacks empirical evidence and undermines all true science. If he is right, then all Biblical miracles could be denied; and on the same basis it could be claimed that the virgin birth of Christ lacks empirical evidence and undermines the sciences of genetics and biology.

He said, while there were witnesses to Lazarus being raised from the dead, there were none at the time of creation and the flood.

He also objects because he feels the doctrine makes God a deceiver of men. One wonders, he asks, "why deity should be so malevolent as to fool us on such interesting matters as much of the history of past events and the possible ages of many things, especially where it is the sort of delusion from which we poor mortals cannot escape."

If the Scriptures are our standard for truth, then creation with appearances is glorious and not deceptive.

Jesus' glory was revealed in this miracle because it involved a supernatural and sudden creation of a complex entity apart from natural processes. This, if it can be believed, is how the Lord Jesus Christ revealed his glory in the creation of the world.

Creationists believe that this doctrine is very logical and scriptural, and it seems to be a theme found throughout the scriptures regarding his way of doing things supernaturally.

Theistic evolution tries to combine both creation and evolution.

Theistic Evolution

Creationists suggest that there were times when a person viewing evolution and scripture would take sides and be an evolutionist or anti-evolutionist. But later came compromises in which the sides were combined, and the result was theistic evolution. People believed here that both The Bible and evolution could be true and could be reconciled. They believed evolution was true and God did it, and it did not come about through mechanistic means as Darwin and other evolutionists proposed. Bolton Davidheiser, in his tape, And God Created, says this is not scriptural. He supports his statement by further explaining the theistic evolutionist's stand. He says that most of these people generally say, in reference to the scriptural account of Adam's being created from the dust of the ground, that the claim is true but leave out the sudden supernatural act of God. Instead, they say that creation went through a long process: that Adam was not made directly from the dust of the ground. He came from the dust of the ground through a long animal ancestry.

Davidheiser suggests another factor God put in the scripture to frustrate the above belief: was the details of Eve's creation. If Adam came from the dust, through a long animal ancestry, why then was there no mate for him among all the creatures of the earth; and why did God have to form Eve from a rib in Adam's side. If there had been no woman for Adam, what would his mother have been like--if he had come through animal ancestry from the dust of the ground?

Creationists also conclude that there is another important factor in Christian faith to be considered: the doctrinal implication of evolution. Some scientists who write textbooks or who write for the general public say, as Darwin did, that evolution need not interfere with one's religion--they are reconciliable. (Transcendental Meditation and other Eastern religions involving meditation do also. They, too, are based on evolution, Karma, and survival of the fittest. They give a system of good works and right behavior as a way to God or peace, as well as following the meditation techniques.) God says salvation is a free gift and there is nothing we can do to earn it. (Ephesians 2:8,9) Because we love him, we keep his commandments --not to come back as a higher form in the next life. It says, "And it is appointed unto men once to die, but after this the judgment." (Hebrews 9:27)

Do these scientists mean by this reconcilibility that there is a First Cause which is beyond the range of scientific investigation? Because of this first cause, that one calls God, there should be no interference with religion. Some say that those who make an issue of this just don't understand.

Progressive Creationism (Threshold Evolution) is thought of by some as a type of Theistic Evolution. Those who do not mind it being called that suggests it is acceptable if it is done without adverse criticism. They say it differs from theistic evolution in that there were several creative acts. (Theistic Evolution says all of evolution is true, but divinely directed.) Progressive creationists say there were acts of God along the way, and evolution was unable to proceed all the way. Certain forms were created and then evolved, but could not do so fully; so God had to create again. He repeated creative acts so as to lift animals over a threshold to a level where they could begin evolving again to more diversification. We might wonder how many creative acts there were. How fat apart were they, and how much evolution was between them?[14]

Day-Age Theory

The day-age theory is the idea that the days of creation were not literal twenty-four hour days, but instead, long periods of time. The theory of evolution assumes vast periods of time for chance to accomplish its magnificent results in the world of living things.

But when creationists challenge the above by asking,"Why is it that nothing is evolving today?" evolutionists say that anything, no matter how improbable, may have happened given a long period of time in the past.

Creationists reason that according to the Law of Thermodynamics and the Principle of Entropy, any ordered system through the passage of time will disintegrate and become more disordered. High level energy will dissipate into low level energy so that its force and power for useful work can never be regained.

Any effort to contradict basic laws of the universe can only be frustrated by the addition of time.[15] Creationists view time as a factor in their theory more than in the evolutionary theory.

They suggest that the day-age theory is not true; however, at the time of and during the six days of creation, the Second Law of Thermodynamics was not in effect, according to the Bible.

If, during the days of creation, the Second Law was not in effect, it would not matter how long the days were according to the principle of entropy, because that principle was not yet in effect, remark the creationists.

In further explaining the day-age theory, some Christians tell us that it is possible to give God time to stretch creation days so that each day was hundreds of millions of years long. Each day, that is, was really an age. Days (and their length) can be determined by reading the first chapter of Genesis. Genesis 1:14 says God created the lights to divide the days from the nights and also to serve for signs or seasons, days or years. Well, if days are ages, then what is a year? If a day is an age, then what is a night? So, if we begin to interpret the word "day," the passage loses its meaning, and there's no sense to it.

It is true says Whitcomb,
that in other places in the Bible the word day can mean a long indefinite length of time as the day of the Lord. But when day is connected with a number in the scriptures, it always means a twenty-four hour period--for instance, the second day, the fourth day, the sixth day. . . . Whenever the word day is connected with a qualifying phrase, evening and morning, we find a technical Hebrew expression that speaks of the rotation of the earth's axis in reference to a fixed light source passing through a night-day cycle."[16]

Genesis 1 is explained by Exodus 20:9-11, "Six days shalt thou labor and do thy work. . . For in six days the Lord made heaven and earth, the sea, and all that in them is. . ."

Saint Peter explained the significance of creation events saying, "that one day with the Lord is as a thousand years and a thousand years as a day" (II Peter 3:18). He would not mean one long period of time is with the Lord as a thousand years. His intent was that in a brief period of time God can, in a literal day, do what man or nature would take one thousand years to effect; because he is not limited by natural processes and time, according to the creationists.

The Gap Theory

Is there a great length of time where fossils and remains of plants and animals can be placed?

For 100 years or more, Christian theologians have studied the book of Genesis in the light of modern scientific theories concerning the origin of this world. Many have concluded that the verses permit the interpretation that before the creation event described in Genesis 1:1--that of the earth in its present form and of living things as we see them now--in six literal, 24-hour days, there was a previous world that God originally created in the manner of Genesis 1:1 which, because of the fall of Satan, was judged by God, destroyed and covered with water. It thus became "without form and void" engulfed in the darkness which covered "the face of the deep." Consequently, prehistoric animals such as dinosaurs (and perhaps even prehistoric man) are to be traced back to this previous world. This theory, which has been widely popularized in the footnotes of the Scofield Reference Bible, has been accepted by many Christian students as the answer to the conflict between the Bible, on the one hand, and the claims of evolutionary science, on the other.[17]

The gap theory has been widely accepted among Christians, especially since the early 19th century when Dr. Thomas Chalmers of Scotland popularized this interpretation, presumably with the motive of harmonizing the essential account of creation with the vast periods of earth history demanded by uniformitarian geologists. In a footnote Whitcomb says that the gap theory had been advocated in one form or another for centuries. It was made popular by Chalmers when he attempted to incorporate George Cuvier's concept of geologic catastrophism into a biblical framework.

The Gap Theory may provide a way to explain the long periods of time required for evolution to occur or offer an explanation of the geologic time table (without implying that all who believe in the Gap Theory also believe in evolution); however, it omits the supernatural, and seems somewhat of a compromise. According to the creationist, God does not require time for creation. He creates instantaneously "by the word of His power." He could create the earth with an appearance of age or with the appearance that it had been here many years. He created it ready for man's use as he created full-grown fruit trees, full-grown animals, and Adam and Eve full-grown. They had an appearance of age.

Evolutionary Origins

Creationists have a problem in finding some means or mechanism that could possibly produce the changes of plant and animal forms or kinds required by the doctrine of evolution. But to date they believe no satisfactory mechanism has been found. There have been suggestions, but there have been difficulties with any mechanism. The most seriously considered mechanism today is the accumulation of successive random mutations.

As mentioned previously, mutations are changes in the organism. Evolution says that occasionally there will occur mutations that aid in the survival of the organism in its environment; and supposedly, a succession of such mutations produces evolution. No known mutation is beneficial if the entire effect is taken into account. For instance, it is a known fact that some mutations reduce the viability of the organism.

The majority of mutations are harmful, and this fact most evolutionists will agree to; yet they hold to the belief that some mutations--although a very small proportion--are beneficial and an accumulation of a succession of these mutations results in evolution. It is supposed that one life form over long periods of time will change into another more complex form.

In other words, changes produced by mutations are in effect supporting evolution, because there may be only several beneficial mutations in a thousand. Furthermore, gene mutations do result in the appearance of new traits, even in some degenerative or modified variational characteristics of already existing traits.

Even supposing that by the smallest chance one organism might "evolve" by this improbable process is sufficient. Organisms depend on a balance in nature which must also be considered. If great emergent changes did take place, it would involve the entire biosphere. The living world, including producers, decomposers, herbivors, and carnivors, is highly complex and interrelated. For example, several plant types depend on insects for pollination. Mutations in all these organisms filling the various niches could have been timed so that the entire ecosystem would evolve as a unit.

Artificial mutations produced by man-made changes in the environment give some clues to problems associated with mutations. For instance, a large dosage of x-rays can produce mutations in a living organism: some are harmful. Using a spray of DDT to kill flies is effective, but after it has been used a long time, some of the flies resist and do not die. Some of them are probably naturally mutated and can survive in the presence of DDT, which acts as a naturally selective agent so that the mutant flies survive and reproduce.

Is the net result beneficial beyond the state of the original house-fly? There is evidence that this type of mutation has a detrimental by-product, because the fly population soon returns to the predominantly non-resistant type again after the spraying has been discontinued. The small number of flies of the original strain which were not killed produced more offspring than the mutant (resistant) type. This resistance to the effects of the chemical could be beneficial to their offspring, but not to man.

High-energy radiation from man-made nuclear explosions or even from naturally occurring nuclear processes on the sun cause harmful mutations. There are international treaties banning these tests, based on this scientific knowledge. These mutations are not expected by anyone to be beneficial; but, even so, mutations are highly beneficial in many environments and certainly must have been a factor in variation if all forms are genetically related.

Establishing a New Trait

For years evolutionary critics have found it difficult for a new and useful structure of a plant or animal to be established by the process of natural selection or that of establishing a new trait. The mechanism of natural selection is claimed by proponents that an animal or plant succeeds because of having useful structures or organs. This seems reasonable. It is supposed that if organs become better, the organism succeeds better and has more offspring--the beginning of a new and improved kind.

Of course, the difficulty lies in that, if the new structure is begun but not developed well enough to be functional, the structure is more of a hindrance than a help. For example, wings not large enough to use in flight would hinder a running bird. The wing stubs might even enable predators to hold the bird more readily.

The vertebrate is a highly complex organism. The idea that it developed through accumulation of chance changes which occurred one at a time may be illustrated thusly: an animal without eyes could have developed eyes. The various stages of development of the eye (lens, retina, nerves, brain to interpret the message) would occur as new additions in the natural selection process, which would not help an animal in the struggle for existence until the animal could actually see with the eye.

Creationists believe natural selection does occur, but they explain the results of the natural selection processes in a different way than evolutionists do. The way that chromosomes, genes, DNA codons, are arranged is so complex and well-structured that it is hard to imagine any improvement from chance, accidental changes.

In nature we see occasional defective plants and animals. Natural selection processes function to remove defective and abnormal organisms before they reproduce. So, in essence, natural selection sets a limit, and a standard is maintained from one generation to the next.

Evolutionists do have the problem of accounting for life coming from non-living matter and evolving to the highly complex and unique human being. In the question of origins, some evolutionists believe that life's origin came about through mechanistic processes following physical laws. The evolutionists have concluded that the tens of millions of fossils in the museums of the world give evidence to the fact that evolution does occur.

The mechanistic theory of life's origin does not violate the Law of Biogenesis that life only comes from life and there is no exception to this law.

One widely accepted evolutionary theory on origins is based on the Russian scientist Oparin's theory.

Oparin's Theory

Barns feels evolutionary dogma is causing degradation in science and divesting America of the unique scientific advantage that a nation which trusts in God has over one that does not.

He illustrates with the example that many of our scientists have accepted Oparin's theory of the origin of life which, he believes, is a dilemma. This states that at one time the earth's atmosphere consisted of methane ammonia, hydrogen, and water vapor, and that lightning caused chemical reactions that produced organic compounds on the top of the ocean and that this eventually emerged into living things.[18] This is probably one of the most widely accepted theories among scientists who believe in evolution.

The creationists contend that, given the conditions Oparin postulated, if the sun shone then as now, the ultraviolet radiation through the atmosphere would have been lethal because there would have been no oxygen to absorb it. We have oxygen in the form of ozone to protect us from that lethal radiation now. The Nobel Prize winner, Harold Urey, and Standlee Miller tried to duplicate the Oparin process and were successful.

Oparin said the original atmosphere could not have oxygen because evolution of life could not occur in oxygen. He postulated ammonia, methane, hydrogen, and water vapor to have composed the atmosphere.

At Professor Urey's urging, Standlee Miller put methane, ammonia, hydrogen, and water vapor in some glassware and sent electric sparks through this gas for a period of a week. He left out the ultraviolet light. What was produced were some amino acids which he reasoned were a constituent of proteins, proteins of living organisms!

Life in a Test Tube

The following was taken from a letter from Darwin to a friend: "It is mere rubbish thinking at present of the origin of life. One might as well think of the origin of matter."

Efforts and probes are used in the hopes of finding the origin of life or discovering whether life exists elsewhere in the universe. The scientific evidence, so far, is that life is a unique characteristic of the planet earth--it is not found anywhere else in the universe.

Even if life is produced in a test tube, still we wonder where the universe comes from. What caused it to come into existence? Even theories of the origin of matter today imply that the universe came into being suddenly, all at once. Physics books refer to the "creative" event that brought about the one hundred elements we know something about today. There are indications, according to the creationists, that elements appeared suddenly and at one time or one moment. The scientific evidence suggests that the origin of living things lies in a single living source, which itself arose from inanimate matter. Emergence of each life form is believed to have occurred from an ancestral form by slow, gradual change. There were transitional forms linking all categories, with no systematic gaps.

In past generations all people have pondered the question of origins. The Egyptians believed life originated in the Nile River. Greeks believed life came from the mud. Then what of mud and water? Where did they come from? Both peoples believed life came from something already in existence.

No group of people on earth, other than the Hebrews, tells us of the origin of matter.

Darwin would not talk of where life came from. He said that all forms of life came from eight or nine basic kinds. Plants came from four kinds of life, and animals from four basic forms. But to ask Darwin where these original forms themselves came from proved unproductive. He did say once,

> There is grandeur in this view of life, with its several powers, having been originally breathed by the Creator into a few forms or one.

> The problem of problems in science today is the origin of life: We must solve the problem of the origin of matter, and then the origin of man. Well, where did life begin, and how can we know things concerning the origin of life? Is there any scientific way of studying this problem, or will it continue to remain the problem of problems?[19]

> Dean further says, "Contrary to popular belief, life has not been created in a test tube."[20]

Yet, it has been reported in the various news media that certain scientists have produced life; but it is also interesting to note that these scientists have denied that they created life as we know it. It is true that some life forms have been produced, but one must note that in the forms so produced, enzymes that came from living organisms had to be introduced. When these living enzymes were placed in the test tube, it was possible to produce some forms of life. But, again, this is merely another instance of life coming from pre-existing life. That it is only possible to produce life in a test tube when an enzyme which comes from some other living organism is introduced into that test tube indicates, perhaps, that life came from some other source of life.

> Herswitz declares, in a chapter entitled "Origin of Life," Many scientists have reported from time to time they have created or produced life in a test tube. What they fail to report is that they, in the production of this life, take substances out of living cells and put them in the test tube, and then they say, We have created life in a test tube.[21]

It may be argued that the virus DNA produced is not totally man-made, because several of the key ingredients used in its production are enzymes obtained only from living forms. But, in principle, these bacterial enzymes, just like the virus DNA, also could be made under the non-living conditions of the test tube. One chemist said that, "If science produced life in a test tube, it would only show that at last man will be able to do what God could do a long time ago." Science suggests that nonintelligence produced intelligence when protein molecules accidentally came together to form life. This does not contradict the fact that naturalistic, mechanistic processes could result from properties inherent in inanimate matter.

Transitions and Evolution

Evolutionists believe that the vertebrate body pattern "evolved" from the invertebrate body pattern. Creationists contend it is hard to imagine how the chief invertebrates, which have an exoskeleton, dorsal circulatory system, central nervous system, and muscles inside the skeleton could "evolve" into the vertebrates whereby the arrangement is reversed with an endoskeleton. The central circulatory system, dorsal nervous system, and muscles are outside the body. Such a transition from the former form to such a different structure--invertebrate to vertebrate--exoskeleton to endoskeleton--seems impossible. The evolutionist challenges the notions directly, contending that the fossil record definitely shows transition in the evolution of primitive mammals from reptilian forms. Further evidence indicates that the Archaeoptery is a perfect transitional form.

Uniqueness of Man

Although man's body has similarities to an animal's body, there are many differences. It is man's human attributes, however, that place him in a group entirely separate from the animals. He is capable of abstract reasoning and uses a complex language. He accumulates knowledge and transmits it to his children. He invents tools. Although some animals may use crude tools such as twigs or sticks, they do not invent and perfect new tools. Man appreciates beauty, composes music, and paints pictures. He is able to weep and to laugh. Evolutionists argue that this is the perfect example of the evolvement of the ultimate creature on earth. Man is not infallible, however: there are questionable circumstances concerning several of his discoveries.

As an example, many have believed evolution to such an extent that they have gone to extremes to be persuasive and final concerning the concept of evolution. There are some who have believed life came from some chance beginning. So, about one hundred years ago, Thomas Huxley, grandfather of Julian Huxley, and Earnest Haeckel, a German biologist, were together working on a project to use material dredged up from the base of the Atlantic Ocean, to attempt to show the very earliest forms of life. They were convinced life had a simple beginning and that it came from a water medium. When some materials were brought back from an expedition in the Atlantic, Haeckel said, "Yes, this is evidence of the first life or some of the first living forms." He invented the term monera, which is still used by some biologists in the textbooks today. After examining the material from the Atlantic, he told his scientific colleagues that he had seen some of the first initial forms or cells that appeared on the earth. Other science colleagues reviewed this and agreed that Haeckel was right. Thomas Huxley was, however, doubtful; but Haeckel insisted that he had found some of the earliest forms. Behind all of this was the technique that had been used to collect the material that was dredged up from the ocean floor. It had been placed in the preserving fluid which was a practice then and still is. Where it was formaldehyde, it is now formalyn. In the formalyn now, we deposit living material--a process known as "pickling." The same thing occurred in Haeckel's experiment material from the ocean bottom. After some ten or fifteen years it was found through chemical analysis that the substance Haeckel thought was material of early life was perhaps the product of chemical reaction of the collected substance and the preserving fluid. It has been suggested that he was actually looking only at precipitated chemical material, and the accumulated remains of more complicated ocean life--shell life that had disintegrated and its decomposed parts that had settled on the ocean bottom--mixed with the precipitated chemical interaction.

In man's early stage (the embryonic), we show some of the early characteristics of our adult life--although in a condensed form. Haeckel was very anxious to say that there was a relationship of complex animals to the human being. So he obrained and sketched embryos of chickens, dogs, and humans. Today biology textbooks are using these drawings and proposing that man in his embryonic form was at one time like the embryo of a dog.

It has been said by some writers that the Piltdown Skull parts and jaws were found because they had been planted to be found. The jaw and teeth had been filed, and the jawbone had been discolored with acid. This was done because someone wanted to pretend that there were fossil materials that could be interpreted as an ancestor of the human being. Piltdown man for 40 years was accepted as an ancestor of the human form. As a result of the analysis of fluoride experimentation, it has been suggested that there was no truth to the idea at all. It was, instead, the jaw of a very recent organism and individual.

THE WORLD OF LIVING THINGS

Classification of Organisms

Kinds of Organisms

The differences between plants and animals seem obvious, plants being green and stationary and animals moving about. Some organisms are hard to classify as plants or animals. A third kingdom, protista, is such a category.

Some early methods of classification of plants and animals follow. (By 1758 as many as 4,236 kinds of animals were recognized, compared to today's one million, and more are added every year.)

The ancient people were not systematic in their classification. Aristotle (387-322 B.C.) was the great Greek philosopher who had a wide range of interests. He divided plants into three groups. Those with soft stems he called herbs. Ones with several woody stems he called shrubs, and those with but a single stem or trunk, he called trees. Animals he divided into land, water, and air dwellers. Other strictly environmental or ecological systems were used.

The Father of Botany, Theophrastus (372-287 B.C.) was the author of the first known botanical work. It was a volume set entitled, <u>On the History of Plants</u>. He studied structures of stems and leaves of plants and grouped them according to the likenesses he found in them. What Theophrastus attempted to do in his classification of organisms is called taxonomy.

In the 17th and 18th centuries scholars began compiling long lists of known plants and animals. Tree plant lists were called herbals, and the animals were beastiories.

John Ray (1627-1705) offered the first clear concept of species, which he defined as offspring of similar parents. But until the time of the Swedish biologist Carolus Linnaeus, plants and animals were haphazardly described and classified.

Though often in ill health and easily overcome with weariness, Linnaeus nevertheless accomplished more than any other biologist in laying the foundation for an orderly system in the study of nature. He was given the honor of full professorship at the University of Uppsala, Sweden, in 1741, when he was only thirty-four years old. He taught botany, natural history, mineralogy, pharmacy, chemistry, and dietetics. He had an amazing collection of plants and animals which his rivals credited to his success as a lecturer. He used them as illustrations.

Those who heard his lectures on the introduction to Systema Natura on God, man, the creation, and nature were more stirred by them than by the best sermon. As his student Sven Von Hardon said,

> If he spoke on the work of the creation and his Majesty God the Creator, reverence, admiration, and awe were painted on all faces.[22]

Linnaeus' outlook may best be summarized by the following quote from Norak Gourlie's The Prince of Botanists:

> The further a man goes the more does he become obliged to admire and praise him. From this we see how far the knowledge of nature leads us to more theology itself and how completely it depicts for us the Creator's magnificant work--the marvelous mountains as a proof of His might, the plants a witness of His skill, the animals an example of His providence. The whole nature confirms that He is wise and the entire world that its Creator is a divine and almighty Lord.[23]

The System of Classification

Linnaeus classified all living things as plants or animals and termed these classifications as kingdoms. In each kingdom are natural structural classifications called phyla. In turn, phyla show distinct groups called classes; classes are divided into orders; orders are composed of families; families are composed of different genera; and genera comprise species. There are often subdivisions of these classifications.

Usually the higher levels of classification are the easiest to distinguish. At the species level is where problems in classification arise. No definition of species satisfies all taxonomists. Usually a species satisfies all taxonomists. Usually a species is most commonly defined as a group of organisms that closely resemble each other and freely interbreed among themselves without, under ordinary circumstances, breeding with other groups. There are cases where organisms classed as different species do interbreed when ranges overlap. An example is the flicker (woodpecker). The ones from the East have yellow feathers, and the western ones have red-orange. They crossbreed freely in areas where the ranges overlap. Linnaeus used the universal and unchanging Latin language which proved descriptive and precise and served for identification of plants and animals. The genus of plants always begins with a capital letter, Rosa for the roses. A genus consists of many kinds of plants or animals, each of which has characteristics possessed in common with each other. The name of the species is written in small letters. Both names are written in italics in printed descriptions and underlined in manuscripts. For example, we have Rosa chinesis, and it means the rose from China.

The basis of Linnaeus' classification was structural similarity usually the bone structure of animals. This system gives a great deal of information, because we can be sure that all animals placed in the same genus are physically closely similar in structure.

Linnaeus, in his book, Systema Natura, recognized wide variation and even hybridization. He often refused to list certain plants as species because he correctly recognized that they were naturally occurring hybrids. He also recognized that much of the variation in the specimens that he collected personally or those that were sent to him was due to effects of hybridization within the species or varietal hybrids.

We can find an amazing variation in dogs, though we know they have resulted in hybridization and breeding. They are also interfertile except for great differences in size, such as the doberman pinscher. But these we do crossbreed with other dogs and thus contribute to the common gene pool.

In human beings there is some wide variation within the species, homo sapiens. They are all of one species, since they are all capable of interbreeding.

In regard to the wide variation in man, creationists have the view that. . .

> All the potential physical characteristics of all peoples were already present (by creation) in the genetic systems of the survivors of the flood. Most of these characteristics were not outwardly expressed, however, as long as the entire population lived together and intermarried with a free flow of genes. Only the "dominant" characteristics would be apparent, with the "recessive" characteristics being merely carried in the genes.[24]

In scientific studies in genetics it has been shown that there are three factors that can cause rapid appearance of new varieties in a species: (1) a rapidly changing environment, (2) small population, and (3) inbreeding. Genetic variations in a gene pool can arise by sexual recombination and mutations. Under these conditions, characteristics that were recessive in the large population quickly become dominant in small populations.

When truly distinctive species are crossed with each other, usually some sterility results. An example of this is a cross between the horse and the donkey which results in a mule. The mule is a hybrid between the two species, is sterile, and so does not constitute a new species. In birds, such controlled hybridization is difficult. So tests on bird species have only recently been made. With plants much more has been done along this line, and distinctive species usually show a considerable amount of sterility, both in the production of fertile ovules and effective pollen.

Investigations of forms such as the fungi, where such studies can be rather easily made, have shown that this sterility results from a major difference in the arrangement of the DNA sequence of nucleotides. In higher plants, where studies have been done, a study of the chromosomes shows great differences in the shape of certain pairs--in absolute size and often in chromosome number.

It is important to distinguish between variation and mutation. Evolutionists believe that variation is essentially due to mutation, since from this viewpoint all species of a genus, for instance, were originally alike and diverged as mutants that were advantageous and thus incorporated into the DNA pattern. It is clear, though, that some differences occur too infrequently to be considered as mutational, and others occur in regularly observed patterns of Mendelian inheritance. Creationists say that this situation has always been true and that the various categories were created with a viability potential, making it more readily possible for them to survive in the various environmental situations to which they are exposed. To put it a different way, there are latent recessive genes that later become expressed.

Because of the difficulty in classifying some organisms, certain ones are placed in a separate kingdom, Protista. Most organisms are distinctly plants or animals, but those for which such classification is difficult are now classed in the kingdom Protista. Euglena is an example of a microscopic organism that behaves like an animal in some ways, yet has chlorophyll which manufactures nutrients like a plant. The organisms usually included in Protista are one-celled or simply organized beings in contrast to the multicellular and more complex plants and animals. Most scientists agree that bacteria and blue-green algae which some scientists classify as the monera, should be included in Protista. The problem is where to stop. Should all algae and bacteria be included? Algae include blue-green forms and giant sea weeds.

Classification is largely a matter of opinion, and there is frequent disagreement among taxonomists. They do not disagree about the characteristics of the organisms being considered but about the categories used in dividing the characteristics into separate classifications. Nature itself is not classified; man has a system of classification of the natural world as he sees it.

SMALL PLANTS AND ANIMALS

The invention of the microscope opened a new world to man. But study has shown that the size of these organisms is by no means a measure of their importance, for they make up by their numbers what they lack in diameter. Furthermore, some of these one-celled creatures are veritable chemical laboratories, synthesizing such important substances as vinegar and alcohol, and even secreting toxins able to kill people. Yet much of the animal world is dependent upon these microscipic plants and animals for food.

Among these plants are fungi, plants that contain no chlorophyll, which is the complex chemical that in the presence of light breaks down water and carbon dioxide and recombines their atoms to make carbohydrates (process of photosynthesis). Slime molds have characteristics of plants and animals, and so are classified into a third kingdom, Protista, with many other organisms.

Another of the small plants are molds which reproduce by spores. There are blue-green molds, bread molds, and water molds.

Listed with other fungi are the parasitic mildews, rusts, smuts, and yeasts. Mushrooms and toadstools are fleshy fungi that contain gills which produce spores. Puffballs have gills inside and can be eaten if gathered before the spores mature.

Some funguses cause skin infections, such as athlete's foot and ringworm.

Some fungi are beneficial in producing medicine. Certain molds give off substances that control or kill bacteria. These are antibiotics. Others are good for food.

It is questioned whether viruses are living or non-living. They reproduce when inside a living cell. It is a debatable question, since viruses behave as living things when inside a cell but assume a crystalline, inactive form outside cells. Some evolutionists feel these are evidences of the evolution of life from non-living matter. The problem with assuming viruses to be an intermediate step is that they cannot multiply outside living protoplasm; and being parasitic, how could viruses have existed before the very cells they must have to exist? Because of these facts, evolutionists believe viruses represent cells that degenerated.

There are three classes of viruses--bacteriophage that live in bacteria; plant viruses that live in cells of plants and humans, and animal viruses that attack human cells. In these classes, they only attack certain kinds of cells, as cold viruses only attack the respiratory tract.

Bacteria are one-celled plants and multiply by cell division, or sexual reproduction or spores. There are three shapes or forms of bacteria--spirillium (corkscrew shape), bacillus (rod-shaped), and coccus (tiny spheres that are single or in pairs or chains or colonies).

Other forms of micro-organisms are rickesettsiae, globular rod-shaped organisms that reproduce by fission but require a host like viruses. They cause diseases such as Rocky Mountain fever, trench fever, Q-fever,etc.

Vectors are living things that carry disease-causing microorganisms from one host to another.

Spirochetes are long, slender coiled or spiraled organisms which are between bacteria and protozoa in size. They reproduce by fission. One species causes syphilis.

Microorganisms can be controlled through immunization or resistance through natural or acquired immunities, or white blood cells. Several are used as antibiotics and chemotherapeutic agents (chemicals that fight disease).

Algae are also very small plants, the simplest green plant-like organisms, and belong in the phylum Thallophyte. Some taxonomists no longer classify them as a plant phylum but classify them into five phyla of the kingdom Protista.

Taxonomists do not always agree about how organisms should be classified. The higher plants are the Metaphyta, and higher animals are the Metazoa, according to some classifiers. Evolutionists cannot always agree about the course of evolution because their hypotheses regarding the past cannot be examined directly in the laboratory or the field.

FOOTNOTES

[9]George Howe, And God Created--Louis Pasteur, Vol. 1, Tape 1. (California: Creation Life Publishers, 1973).

[10]Ibid.

[11]John C. Whitcomb, And God Created--Creation of the World and Methods of the Creator, Vol. 3, Tape 1. (California: Creation Life Publishers, 1973).

[12]Ibid.

[13]John C. Whitcomb, Jr., The Early Earth (Michigan: Baker Book House, 1977), pp. 30-31.

[14]Bolton Davidheiser, And God Created--Theistic Evolution, Vol. 3, Tape 2. (California: Creation Life Publishers, 1973).

[15]John C. Whitcomb, And God Created--Days of Creation and The Gap Theory, Vol. 2, Tape 2. (California: Creation Life Publishers, 1973).

[16]Ibid.

[17]"Whitcomb, op. cit., Vol. 2, Tape 2."

[18]"Barnes, op. cit., Vol. 3, Tape 2."

[19]"Dean, op. cit., Vol. 4, Tape 2."

[20]Ibid.

[21]Ibid.

[22]John N. Moore and Harold S. Slusher, Biology A Search for Order in Complixity (Michigan: Zondervan Corporation, 1976), p. 142.

[23]Ibid., pp. 142-3.

[24]Alber Hyma and Mary Stanton, Streams of Civilization (California: Creation Life Publishers, 1976), p. 33.

CHAPTER III

ZOOLOGY

DEFINITION

Zoology is that branch of biology that deals with the study of animals.

DOCTRINES OF BIOLOGICAL CHANGE

There are more than one million species of animals. The evolution theory contends that the first living cell "evolved" into complex multicellular forms of life. These gave rise to all forms of animals without backbones; in turn, these animals "evolved" into animals with backbones. Fish then "evolved" into amphibia, amphibia into reptiles, reptiles into birds and mammals, early mammals into primates; and finally, primates "evolved" into man.

Both evolutionists and creationists have prepared their separate doctrines to explain the same facts of biological change. The facts of biological change brought out in breeding studies and in differences in fossil remains of different strata are observable, but interpretations of such facts by evolutionists and creationists are influenced by the doctrines of biological change to which they subscribe.

GEOLOGICAL EVIDENCE AND HISTORY

History of Evolutionary Theory

George L. Buffon (1707-1788) was a French naturalist and was a very influential man of his time in developing the geological theory. He was against the idea of a universal deluge. He believed the earth's history was divided into seven epochs, and this became the basis for the day-age theory. This theory, as mentioned earlier in this paper, maintains the Creation days of Genesis were not twenty-four hours in length but millions of years long. Carolous Linnaeus (1707-1778) believed that all species were created and changed little or not at all since creation. Later he saw the possibility of great variations. The father of Charles Darwin, Erasmus Darwin, was an evolutionist who believed in survival of the fittest. Geoffrey St. Hilaire (1772-1844) was an evolutionist who thought it was true that sudden changes occurred in organisms. This idea foreshadowed Hugo de Vries' mutation theory. George Cuvier (1769-1832) believed that different fossil beds represented catastrophic destructions with migrations from other regions taking the place of the destroyed forms.

Jean Baptiste de Lamarck (1744-1829) was the first man to produce an organized theory of how presumed changes may have occurred. He said that environments brought about evolutionary changes, in his theory of inheritance of acquired characteristics. An example of this theory is the giraffe's long neck. He thought that as the giraffe ate the leaves of the tree it would stretch its neck a little to reach the highest leaves. He felt that then the slightly lengthened neck was inherited by its offspring who, in turn, stretched its neck a little longer. This evolution supposedly continued to the long-necked giraffe we have today. Lamarck believed that a change of any kind that takes place in an organism can be passed on to its young. This idea of acquired characteristics is not accepted by today's biologists. Although environmentally-caused characteristics cannot be inherited, environment can influence the development of hereditary characteristics. An example of this might be a child born with a musical talent who may not develop it if his environment prevents him from doing so.

Charles Darwin, who wrote (in 1859) the book entitled Origin of the Species, explained the theory he developed while working as a naturalist on an exploration voyage on the H.M.S. Beagle. According to his theory, young organisms may differ from their parents in every direction, and more young are produced than the habitat can support. Individuals struggle against others to exist; only the fittest survive, and the characteristics of the survivors are passed on to the next generation. As a result of this process being repeated several times for many generations, the form of the organism changes and new species result.

Darwin's survival of the fittest would explain the long neck as follows: In each generation of giraffes, the lengths of the necks would vary somewhat.

The ones with longer necks could reach higher to get leaves. The giraffes with the shorter necks would be less vigorous in the struggle for existence and thus would not produce as many offspring, because they could not get as much food as the longer-necked giraffes. The long-necked animals would pass their traits on to their offspring, which would then have longer necks than the generation before. The long necks would have been such an advantage that eventually short-necked giraffes would no longer exist. Natural selection would eventually result in greatly increased lengths of necks.

Later Darwin came to the conclusion that natural selection was not the only mechanism of evolution. He went back to Lamarckism. He thought that the body cells formed particles he termed "gemmules," which reflected the influence of environment. He believed these accumulated in the reproductive organs by means of the blood and determined the characteristics of the next generation. As a result, the genes would change slightly every time a new offspring was born and the changes could be in every direction.

At the time of publication of the _Origin of the Species_, this kind of reasoning seemed valid. Around the turn of the century, however, the work of Gregor Mendel (1822-1884) an Austrian monk, was rediscovered. His work, published in 1866, suggested the genetic laws that he discovered when he did experiments with breeding. Until 1900 Mendel was ignored. In one of these experiments he crossed the tallest and shortest peas in his crop and bred them with their own kind, showing the progeny appeared in definite numbers per phenotype, which in turn were reducible to simple ratios of whole numbers. The laws of Mendel changed the whole concept of inheritance. He proved the fallacy of the Lamarckian inheritance theory. Darwin was said to refuse to read his work.

So, Darwin didn't succeed in accounting for the origin of new characteristics that he believed provided the substance of evolution. A Dutch botanist, Hugo de Vries (1848-1935) proposed the idea that mutations provided the variations upon which selective pressure could act to produce new species. Darwin knew nothing of genetics but de Vries did when it was becoming established as a science. We know today that genes only change by mutation and that this occurs rarely. The mechanisms of mutation, not selection and gemmules, are adequate to bring about major changes that can be inherited and cause an evolution to be a more complex and desirable condition.

In speaking of giraffes, the slight differences in neck lengths is known today to be caused by differences in food or by possession of a different number of the dominant genes that control neck length. Mutations resulting in greater neck lengths never have been observed.

Neo-Darwinism or New Darwinism is the term given to modern evolutionary beliefs because Darwin was the first one to convince the people to believe evolution was true.

In order to prove the doctrine of evolution, proponents would have to show today that one animal kind or form is changing into another kind or form. The fossil records do show that life succession through genetically related organisms has passed through changes from the small to the large, from the less complex to the more complex. The greatest diversity of living forms is found in the most recent rocks; and many living things, including higher plants (and man), have appeared in these fossil records.

Fossil Evidence for Evolution

Many believe the fossil record is the best evidence to support evolution. Yet others believe that fossils can be used to discredit the possibility of major evolutionary change. It is stated by some that the theory must stand or fall according to the fossil evidence, because the fossil record is the only available historical evidence.

Fossils are interpreted as evidences of previous life. The original substance may either be preserved, replaced by some other material, or some fossils may be imprints. In the first category, an example might be insects preserved in amber or frozen plants or animals. Frozen mammoths have been found in Siberia whose flesh was so fresh after thawing that dogs could eat it. Some mammoths have been found with plants still not digested in their stomachs. So it would seem that they were frozen suddenly. It would appear that they had lived in a warm climate because of the plants they were eating when suddenly they were frozen. There must have been a sudden temperature change of 100 degrees fahrenheit. A cause other than a catastrophy can hardly be imagined.

The majority of fossils are of the replacement type. One example is the petrified forest. Usually only the organisms' hard parts, such as skeletons and skulls are replaced with the substance of silica, calcite, and pyrite. Some replacements are in much detail. In others, only general shape is replaced.

The third category includes tracks of crawling animals, worm holes, foot prints, and imprints of other parts of the body of the animal or plant.

In order for a plant or animal to be preserved, it must be buried rapidly (in ice, wax, stone, asphalt, etc.) or else the organism will decompose or be eaten. This is especially true of soft or marine organisms because they degenerate soon after death.

Geologists use fossils to determine the relative age of rock layers. They assume the deepest layers are older in terms of deposition than the overlapping rock, unless the layers have been disturbed. Toward the early part of the nineteenth century, a study of rock layers was made that revealed that the kind of fossils in a given series of layers or formation were mostly the same. William Smith, a land surveyor, had the idea of fixing the relative ages of strata by their fossil contents. This was accomplished by using a particular species of an animal as an index of a rock formation. In order to be a good index fossil, the animal must have existed for a short period of time (which would make it characteristic of a set of strata) and it must be abundant and have a wide geographic distribution. It was assumed that the rock layers or formations were laid down over the earth in the same order as in England. For example, English and Silurian rocks were believed to have been laid down at the same time. Animals characteristic of these rocks were believed to have come into being at the early Silurian Period. Also, if any sedimentary rock such as shale, sandstone, or limestone contained these fossils, it was classified as Silurian.

In 1860, after the theory of evolution became popular, it was decided that index animals did not become extinct at the end of each period, but that they changed gradually to other species. This dating system, still used today though slightly altered in form, has many objections to it. Geologists no longer expect to find all the layers they observed in one place to occur every place they go. But they have decided a priori (considered valid independently of observation) that the relative order of all layers extends over the whole earth. By this scheme, geologists believe they can predict when an isolated formation occurred. All that one can tell by the principle used--namely, that the deepest layer is the oldest--is a local deposition order. It does not provide a way of giving a date to a stratum, nor does it tell what time relation it has to formations in other parts of the world.

Creationists hold to the idea, that in the past, there were life zones just as there are today. Creationists pose the question: Why could trilobites not have lived in one life zone at the same time dinosaurs lived in another? In light of present conditions, it is difficult to imagine a time when the earth was dominated by fishes, another time by amphibians, etc. It is true that in a given area where several periods are represented, the deepest layers contain usually the least complex forms. The explanation can be clarified by the fact that the majority of fossil material was laid down by the flood in Noah's time. As the flood waters rose, the simplest forms, being less able to escape, would be buried first. More complex and more mobile forms could move to higher ground. The evolutionists contend this is the greatest flaw in their theory. Most surmise that the more one considers the fossil record, the more laughable the notion of "flood geology" becomes.

The fact that most fossils are fragmentary presents a problem to paleontologists. There are a few rare exceptions to the rule that only the hard parts such as bones and skulls are preserved. These exceptions are the frozen mammoths in the Arctic and insects in amber. The soft parts of most fossils must be deduced. If the fossil is a familiar animal, such as a wolf, it is logical to assume the muscles and internal organs are like those of wolves today. But we cannot know, for instance, his color, the size and shape of ears, type of hair, etc. Paleontologists, because of these types of uncertainties, have been known to make mistakes when making restorations. An example of this is that drawings of the mammoths were different than the entire animal found frozen in the Arctic. Also, mistakes in dinosaurs have been discovered where plates along the spine were not correct. We should be wary of restorations of the past based on the fertile imaginations of paleontologists. It seems to follow that even the so-called fossil men may not have had the facial features textbook artists have drawn. We are not sure of the details of the evolvement of ape-like creatures to modern man. The facial features are soft parts that cannot be determined exactly by man.

Extinct Organisms

Fossil material is interpreted as evidence that many organisms which existed in the past are now extinct. This does not in any way necessitate the conclusion that these extinct forms gave rise to the modern forms. In the majority of cases, some investigators believe, they are forms which lived simultaneously with forms like those of today. Some forms paleontologists do not attempt to restore, because there are no living forms like them today for comparison. Restorations of dinosaurs, however, are attempted even though paleontologists know much of what they restore is what they have imagined. Yet they have bone, skin, and scales to provide a basis for their work. Pictographs of dinosaurs have been discovered on walls of canyons and caves which certainly give some credence to earlier life forms.

The Fossils of Diseased Organisms

There is another problem of interpreting fossils related to their possibly altered structure because of disease or injury of the organism while still alive. Some fossils give evidence that organisms in the past suffered some of the same diseases and injuries of today's forms. For example, the dinosaurs are believed to have had rheumatoid arthritis.

Due to a disease in the Neanderthal man, he was pictured at first as a stoop-shouldered, bent-kneed, hairy, cave man. This was based on a skeleton found at La Chapell, Aux-Saints France. It was later discovered that this particular individual had osteoarthritis that caused his back to be curved like the curved shape in pictures in cavemen restorations. Other Neanderthal skeletons discovered later were probably as upright in posture as people of today. They probably looked like people of today, according to some authors.

Besides diseased individuals, size also is a problem and can be misleading to paleontologists. A farmer knows that animals grow larger if they are well fed than if they were deficient in food, minerals, and vitamins. It is known, too, that temperature and chemical content of water can influence the size of living shellfish and other sea animals. There is evidence that this is true of fossilized animals as well. Some fossils are composed of iron compounds, and some paleontologists believe that the iron contained in the water was the cause of the stunting. Not all iron-containing fossils are small. Sometimes, though, a known species may be small and ferrunginous; but scientists do not know whether they are small because the iron dwarfed them or the organism is just a small variety.

Fossil Classification

A consideration of living species provides another problem for fossil interpretation by paleontologists. For instance, lion and tiger skeletons are so nearly alike that if they were found as fossils, they would be classified as the same species, though as living animals they are separate species. If dog skeletons such as chihuahuas and Saint Bernards were discovered, the varieties would differ greatly from each other. Thus, if found in fossil forms, they would most readily be classified as different species due to this great variation of characteristics. Also, some animals are very different at different ages. For example, the tadpoles and adult frogs or the larva and adults of insects are so much different that it is possible to classify some fossils as different species. Also the sexes of some kinds vary greatly. It is possible that because of some of these differences many fossils are erroneously classified.

Hybrids may be a cause for erroneous classification also. The mule is a sterile hybrid between a male donkey and a female horse. In some cases extinct animals may also have produced sterile species that are classified by paleontologists as species. This could be true in the horse species and the sea urchin series in England.

There has been much written about the evolution of the horse. Just how much evidence there actually is for the horse "series" is difficult to ascertain, because information seems to be unobtainable as to how many specimens of a supposed genus have been found and how complete they are. There are at least twenty-six known genera today. As a result, the horse evolution is a possible variation within describable limits of the horse kind.

Horses of all sizes can be developed through selective breeding within one species. The variety of horse sizes from a midget to the one-ton Clydesdales comes from the same gene pool and can be developed without mutations.

The "horse tree" is more like a bush than a tree, as most of the genera are not believed to be direct ancestors of our present forms. Eohippus, sometimes interpreted as small as a fox, had four toes on its front feet and three on its hind feet and is usually placed at the beginning of the series. The general trend of horse evolution is supposed to be: (1) increased in size from a cat to some larger than existing horses; (2) enlargement and lengthening of the head anterior to the eyes; (3) increased length and mobility of the neck; (4) changes of the premolar and molar teeth from types suited for browsing to types suited for grazing; (5) elongation of the limbs for speedy running; and (6) reduction of the number of toes to one long toe.

The fossils of these horses are widely scattered throughout Europe and North America. There is not a place to be found where they occur in rock layers one above the other; neither is there any sequence that would indicate that the largest developed from the smallest. It is believed that some differences in size may be due to the difference in feed. This can be supported by a case in 1942 where a herd of horses was found in a box canyon in Southern California. Three of the horses were caught and carried out with ropes and pulleys. Due to improper feeding (a lack of food), the horses' backs reached no higher than a table. But later a colt was born to these captives and was well fed. It grew much larger than its parents. Since a difference in size due to the amount of feed is an acquired characteristic, it is not inherited and so does not account for permanent changes in a species.

It is possible that the small size of the horse-like animals was due to poor feed and the side toes could have been lost through mutation. It is even possible that some of the fossils may have been genera not related to the horse at all.

The only way to determine whether or not varying organisms are of the same species is to breed the living organisms. We know, of course, that this cannot be done with extinct organisms. But the evolutionists contend that the "flood geology" concept has to make the creation theory erroneous by indicating that all the various forms and species were created at the same time. More bluntly put, the Equus (modern horse) would have treaded water for a longer time than Mesohyppus (intermediate form), and Hyracotherium, a small ancestral form, would have been a poorer swimmer still. Therefore, they would have been buried in an imaginary "sequence" from which we now dig them up. One evolutionist remarked that from their theory, the higher plants would have had to "flee" to high ground, and hence appear only later in the geological record of the flood. He said, "the more one considers this possibility, the more we can believe in our fossil records."

Fossil Dating Methods

Here is a chart of the geological time table which shows the theoretical geologic ages. (Figure 5) This chart grew out of the principle of uniformity, the doctrine of which Sir Charles Lyell is considered the father. He stated that existing physical processes acting essentially as at present, are sufficient to account for all past changes. Geologists seem to lean to this principle and build on it, rather than on the one held by Cuvier and others that rocks were laid down and fossils formed by great cataclysms and upheavals. This uniformitarian principle is in contrast with the concept held by some scientists that the sedimentary layers and major surface features of the earth came about as a result of one major catastrophe of world-wide proportion, namely the deluge of Noah's time.

Figure 5

THE GEOLOGIC TIME TABLE

ERA	PERIOD	DURATION (Millions of years ago)		BEGINNING DATE
Cenozoic (New life)	Quaternary	75	1	1
	Tertiary		74	75
Mesozoic (Middle life)	Cretaceous	130	60	135
	Jurassic		30	165
	Triassic		40	205
Paleozoic (Ancient life)	Permian	300	25	230
	Carboniferous		50	280
	Devonian		45	325
	Silurian		35	360
	Ordovician		65	425
	Cambrian		80	505
Precambrian (Proterozoic Archeozoic)		1500		3000
Azoic (Without life)		3000		5000

Geologists must accept a certain amount of catastrophism because some features simply could not have been formed by present day processes. Features can be better explained sometimes as having been formed rapidly by great forces than by mild forces acting over great lengths of time.

A good example of this is the Grand Canyon. Many geologists have interpreted that it was cut over millions of years while the land rose, and the elements did their work. The creationists say that it was formed rapidly as water cut through not yet consolidated material which had been deposited by the flood of Noah's time. They reason that their explanation conforms to the principles of hydrodynamics.

The Colorado River channel is both deep and meandering. Another way of describing the doctrine of uniformitarianism is to say that the present is a key to the past. If this is not true (and there are many cases where it does not seem to be), the geologic time table is not valid, according to the creationists. Since geologists are limited to studying what they find during their present lifetime, uniformitarianism might be better stated: <u>the present is a key to the present</u>.

We see that the earth's history is divided into eras, periods, and epochs. Some scientists assume the earth to be 4.5-5 billion years old. They believe the oldest rock to be 3.4 billion and the first life, about 2.7 billion years. The first date came from radioactive dating of a meteroite. The date for the first life came from dating an algae reef.

The Precambrian is divided into two eras--the Archeozoic the Proterozoic. The former is represented by no fossils; the latter, represented by just a few fossils, none of which are undisputed.

Abundant life appears suddenly in the Cambrian period of the Paleozoic (ancient life) era. The Paleozoic era is estimated to have begun 505 million years ago, the Mesozoic (middle life) 300 million years later, and the Cenozoic, which contains the present, 75 million years. Geologists estimate geologic time in terms of millions of years without any fully reliable means to verify their estimations.

The sudden appearance of complex forms was found in the Cambrian. According to the currently popular geological scheme, strata containing fossils were laid down in 13% of the earth's history; whereas during the other 87%, no unquestioned fossils were formed. In spite of this, life is supposed to have been in existence for 60% of the earth's history. It is rather difficult to explain why no fossils are recorded from Pre-cambrian times. If these Ancestors did exist, it is wondered why they leave no trace when complex forms are found abundantly.

The Devonian period is known as the age of fishes. The Mesozoic was known for dinosaurs, toothed birds, and some mammals. Also, one of the most striking things the Mesozoic era is known for is the sudden appearance of flowering plants which are like present day species. These flowering plants have no ancestors in the fossil record. Darwin called them an "abominable mystery" as they are yet today. In general, plant and animal types appear suddenly. The Cenozoic, which is characterized by mammals and man, supposedly came into being two million years ago. Doctor Leakey was known for two important finds of fossils, Homo habilis and zinjanthropus. Supposedly they lived at the same time. The first one is similar to modern man, and the latter is ape-like. When zinjanthropus was first found, he was called "early man," but Homo habilis was found in deeper strata. This fact forced evolutionists to see a contradiction (as the simplest are found in deeper strata) in their idea of "human evolution." So zinjanthropus was moved out of man's direct line of descent.

The basis of this kind of geological "sequence and time" scales is that according to evolution, the first organisms to evolve were the simplest and should be found in lower layers of strata. Then as organisms continue to evolve, the more complex forms were deposited in strata above the lower layers. If all the ages were found in correct sequence, they would be like the diagram. (Figure 5)

However, more than six or eight periods are never found in any one locality, and often there are only two or three. One geologist said that if all the periods were to be found in one locality, the strata would be 95 miles thick. It is considered that the Grand Canyon is an excellent example of strata in correct order. But the only "periods" represented are the Cambrian, Devonian, Carboniferous, and Permian. The Ordovician and Silurian above the Cambrian and the Carboniferous are not present. No periods are represented later than the Permian.

Besides the fact that there is no known place where all the periods are represented, they can also be found in their reverse order. An example of this is the Lewis "overthrust" in Glacier National Park in Montana. This is an example of Precambrian rock that is supposed to be almost 500 million years older than the Cretaceous rock beds over which it rests. Because this is not in keeping with geologic theory, it is postulated that this entire Cambrian and Precambrian formation was thrust physically for tens of miles over the Cretaceous shales upon which it rests. The physical evidence does give support to this hypothesis. There is evidence of disturbance in the layers, except in small areas, which would be expected in an area this size (350 miles wide and six miles thick).

There are also time gaps in many places. Some examples are found in Canada and the United States where a "young" bed can be found resting on a "very old" bed without evidence of an eroded surface between. It would seem certain that if millions of years had gone by between the times the beds were laid down, much erosion would be evident on the surface between them. Another occurrence of difficulty for most geologists is the explanation of the repetition of a layer several times in succession. These things are hard to explain when speaking of uniformitarianism. Geologists have insisted that all the earth's physiographic and geologic features can be explained on the basis of uniformitarianism--the application of present laws and processes, operating in the same manner and at the same rates at present, to the understanding of past deposits and activities. This principle implies there were eons of geologic time in order to account for the tremendous extent and depths of the deposits in the earth's crust. This is an example of man's accumulated knowledge to account for the formations of bare areas being in reverse order. Thus the present may <u>not</u> be a key to the past.

There are two ways the number of years assigned to each era, period, etc., are estimated: (1) by the rate at which evolution is thought to have taken place;

and (2) by fossils used mainly for relative dating in order to determine which formation is older and which younger. Circular reasoning is used in this relative dating. Evolutionists concluded that living things developed from simpler to more complex. Because of this fact there is no area where the whole series is represented, pieces from different areas are arranged together on the assumption that the less complex should be the oldest. The evolutionists came to the decision of the order of the rocks. The order in which they believe the organisms evolved. As an example, fishes are thought to have evolved into amphibians, which evolved into reptiles, and reptiles into birds and mammals.

The shell of an ammonite is an index fossil belonging to the Cretaceous period. Any formation containing this ammonite fossil is considered Cretaceous. At first the fossil was dated by the rock where it was found. So fossils date rocks and rocks date fossils and we have this circular reasoning. Fossils are used for dating in years, only in the sense that estimates are made as to how much time would be required for an amphibian to evolve to a reptile, and a reptile to a mammal, etc. These estimates are important in that they are used to judge whether or not an absolute dating can be correct.

Absolute dating is dating in terms of years and this is based on the decay rate of certain radioactive elements. Moore suggests that "radioactive dating, too, is based on a mixture of fact and assumption."[25] The common dating methods are carbon-14, potassium, argon, rubidium strontium, and uranium-lead. The methods are based on the fact that the elements of carbon-14, rubidium, uranium (usually called or known as the parent elements) decay redioactively and finally change into C_{12}, Ar, Sr, and Pb respectively. These latter elements are the daughter elements. The elapsed time for half of the elements in a given amount to change has been determined. Providing the following assumptions are true, a sample found in the field today can be dated by this method. They are: (1) none of the daughter elements was found in the rock when it was formed; (2) the rate of dacay of the element has stayed constant since the time the rock was formed; (3) all the daughter elements in the rock were taken from the parent element previously in the rock (i.e., there has been no contamination from anything outside the rock). It is possible that some of the parent or daughter elements may have escaped from the rock. These are some of the assumptions involved, the validity of which can be determined. One example is the fact that cosmic radiation can alter decay rates. If the amount of cosmic radiation has varied in the past, these methods of dating would be valid.

Creationists Interpretation of the Fossil Record

There is an alternate interpretation of the fossil record as seen by creationists. They have found other dating methods to indicate a young earth. These methods show the earth to be much younger than the 4.5 billion years now claimed. Many creationists believe the age of the earth to be in the thousands --and not millions or billions--of years. By measuring the amount of meteoric dust in the solar system, as well as the rate at which it is escaping, it seems that it should all have gone a long time before now if the solar system is as old as is claimed. Even the amount of helium in the atmosphere points to a shorter age for the earth. Dating methods by which the earth is interpreted as being young are not accepted by evolutionists, because evolutionists require eons of time as a framework for the doctrine of evolution; and evolution does not have any observable evidential bases.

Many organisms are still in existence today that by all the "rules" of evolution should have become extinct or evolved into something else a long time ago. The shellfish Lingula is found in layers of rock in Ordovician times. There has been no change over an estimated 500 million years. Neither have the Sequoia trees of California changed or evolved. One animal that we might expect to be extinct is the opossum, which has a small brain and is not specialized in its teeth, feet, or legs. Yet it has extended its range from Middle America to New England and California. These and other forms have changed very little from forms like them in the fossil record.

There is still the problem of fossils found in the "wrong" formations. For example, flowering plants that appear suddenly in the Cretaceous period have no ancestors in lower formations. There have even been found fossil pollen grains of the pine family in the

Precambrian at the bottom of the Grand Canyon. According to evolutionists, only much less complex plants were alive when those rocks were formed. On June 1, 1968, William Meister, near Delta, Utah, found a sandal-like print with several trilobites in it. It can be deduced from this that man and trilobites lived simultaneously. But, according to evolutionary thinking at present, trilobites were extinct hundreds of billions of years before man was on the scene. However, since 1968, several other footprints have been found near the same place, and all in the Cambrian strata.

Another amazing find was that of a gigantic man's footprint in a Cretaceous river bed close to Green Rose, Texas. Found in the same bed were brontosaurus tracks. So we can deduce from this that man and dinosaur lived at the same time. Men were not "supposed" to have been in existence in the Cretaceous period . . . or for sixty or seventy million years after that time. The prints were giant in size like those described in the Bible. Humanlike footprints have been found in the Carboniferous at various other sites. But this evidence is ignored by many geologists because it does not fit the evolutionary theory.

Uniformitarian principles would support the idea that organisms died and were buried just as they are today. But usually dead organisms are soon destroyed by scavengers and decomposers. Sometimes floods bury a few dead creatures but in the fossil record there are many localities where tremendous numbers of fossils are found, many more than can be accounted for by present day processes. Due to the excellent preservation and large numbers, a quick burial seems to be the most logical explanation. It is impossible to account for such tremendous numbers of fossils on the basis of the uniformitarian principle. Fossils today are not being deposited in this manner. Creationists believe the huge fossil beds were laid down during the flood of Noah which once covered and destroyed life on the earth.

According to the evolutionary theory, there was a gradual change from single-celled animals to man through several stages. There should be a sequence of fossils showing minute changes from one cell to man, but instead there are gaps, and so divisions of fossil forms in the same major classification groups we use for living organisms is possible. The first fossils were complex organisms. There is no transition within the "kinds." (This word can be closely related to "species," but is not synonymous).

There are two explanations given for gaps in the fossil record. They are: (1) rapid evolution occurred at the points where the different breaks took place. (But this does not go along with the theory of random mutation and selective pressure. What is known about mutations seems to suggest a very slow process.); (2) fossils either were not formed or could have been destroyed at certain breaks. (But since these breaks happen so regularly and correspond to the breaks that occur in the classification of living organisms, this does not seem likely.)

Rather than supporting evolution, the breaks in the known fossil record support the creation of major groups with the possibility of some limited variation within each group. The variations are not great enough to change one family order, classification or phylum into another. This, too, is usually true of the genus; at least, if not the species. Creationists say that the breaks are to be expected if separate kinds were formed initially. They believe discontinuities have existed and still do in the world of life. There is evidence of discontinuities in the fossil record, which means that creationists have confirmation of a major prediction based upon the "Genesis Account of Creation."

Radiometric Dating

By radioactivity we mean that there are atoms such as uranium and others that are blowing up and decomposing. Uranium particularly does this. It gives off a high speed alpha particle which slows down, stops, and becomes a helium atom. So uranium does not just disappear, but changes to another atom and leaves behind an ash product of helium plus the new atom that is there. This is done at a rate called a half life. Half life means that of any given amount (regardless of the starting amount) in a definite time, only half of that amount will be there, and in definite time again only half of that amount will be there. For example, suppose the day before yesterday we had 32 pounds of something. The half life of this object is one day. If two days ago there were 32 pounds of it, then yesterday 16 pounds, then today 8 pounds, tomorrow I will have 4 pounds, etc. Regardless of the amount you start with, you will have half of it left in whatever the time constant is. This time constant, where you have just half of it left, is called a half life.

Uranium 238 has a long half life. If all of the atoms on earth remain--and the atoms of Uranium 238 that are now on earth remain, in four and one-half billion years only half of that many will still be here. Scientists have thus calculated that it takes four and one-half billion years for half of the uranium of mass 238 to change into a new type of atom. We call that change radioactive decay, and because of it helium is produced.

There are two ways to tell age by means of radioactive decay. The half life is a constant that can be used when working with uranium, thorium, and carbon 14. Scientists in the laboratory can do little to speed up or slow down this time constant. In principle, this is an apparently constant thing regardless of reasonable surroundings (pressure and temperature conditions). This appears to be a good clocking technique. We only need to know that the rate at which half of it is gone is a constant. We can do clocking work with it by one of two methods:

1. Knowing the original amount and the amount we have today, we can calculate by half life how much time it took to reduce the original amount to the amount we have at present.

2. As the substance reduces from the original amount to the present amount, radioactive ashes are produced; and the longer the process has been going on, the higher the pile of ashes. By measuring the original height of the ash pile and its present height and how high the ash pile is building up, you could determine how long the process has been going on.

For instance, we might now have a six-inch-high candle that is burning, which is analogous to measuring the present amount of uranium, or any other radioactive species one may use in clocking. The six-inch candle is burning at the rate of one inch per hour like the radioactive species is decaying. We know and can measure the half life. So we ask the question, how long has the candle been burning? (It is now six inches high burning at one inch every hour.) The height being determined, this can be computed. We know the starting conditions, not only the rate but also when it was calibrated. We have determined the original height of the candle and the initial starting amounts of radioactivity. After determining how much carbon dioxide was given off by the candle and how much carbon dioxide was in the closed atmosphere when the candle started burning, we can now determine how much the candle produced. This is true, as well, of the ash product helium in the radioactive dating process.

Do we have any guidelines, reasonable boundaries, as to the assumptions in radioactive dating. It appears we do. The measurement of time by study of the continuous breakdown of radioactive elements has had great impact on science and philosophy. We have learned that the naturally occuring radioactive elements are constantly decreasing in abundance, and this phenomenon forces upon us a new realization. It demands a creation of these elements (and therefore, of all elements) at some definite time in the not-too-distant past.

So, if these elements are breaking down continually, and if this deterioration started "way back" then, we might ask how much would we have today if they have forever been breaking down?

In a like way the decay of all radioactive elements would have been completed in the years past. But, the fact that the decay processes are still going on eliminates the possibility that they have ever been going on forever. Science seeks an answer to origin but there is no scientific way to produce it. So the origin of the universe must have taken place through processes that are not going on today--namely, creation by the spoken word of God, according to the creationists.

UNIFORMITARIANISM AND CATASTROPHISM AND THE FLOOD

Uniformitarianism, as mentioned earlier, depends on the idea that present day geologic processes acting more or less uniformly as they do at present, can explain how all these rocks were slowly formed over long ages in the past. Catastrophism is based on the idea that a world-wide catastrophe formed most of these rocks. The most important part of the catastrophe was a great deluge of water; but, also, with it came gigantic earth movements, volcanic eruptions and other violent phenomena. Catastrophists believe that present-day processes could never have accounted for the great mountains in the world, or the great expanses and thicknesses of sedimentary rocks, or other numerous features of the earth, such as the great masses of fossilized plants and animals which were buried in the rocks in all parts of the world. They feel only the flood can account for these events.

The Uniformitarian concept, however, believes it is more scientific to assume the earth's processes have been more or less as they are today. (The present is a key to the past). Also, it would be necessary to have long ages in order for evolutionary processes to function adequately. Creationists support the fact of uniformity since the flood, for the Bible records:

> While the earth remaineth seedtime and harvest, cold and heat and summer or winter and day and night shall not cease. Yet, at the time of the flood, God miraculously intervened in the processes of the past. These processes were different than today's--the climate previously having no rain. He speeded up the processes through the catastrophe of the flood. His intervention ended this uniformity during the time of the flood, because great cataclismic and catastrophic processes or events were brought into play and the processes were therefore, not as they are today. (Gen. 8:22)

The possibility of catastrophism as an explanation of the earth's geological formation has been arbitrarily rejected by most geologists for well over a hundred years. They have insisted that the earth's physical and geological features are explained by uniformitarianism.[26]

With the concept of uniform natural law throughout the ages since creation, the Biblical cosmologist has no quarrel. The two laws of thermodynamics in particular, which constitute the basic framework within which all natural processes operate, have undoubtedly been in effect since that time, including the flood period.[27]

The uniformity of process rates is a different matter. All natural processes operate within the framework of the two laws, which means that not only is every process a conservative one but that it is also a decay process. However, the Second Law says nothing of the rate of decay; this depends on the relative influence of the various parameters controlling the process. And because any of these may easily vary in time and space, it is easily seen that all process rates are subject to change. So then, no such rate can be used with full confidence as a geochronometer, either qualitatively or quantitatively.

Historical geologists in recent years are widely recognizing that geological process rates are subject to wide variation from zero to catastrophic intensities. The actual formations are often of such character as to preclude explanation by anything other than high intensities of sedimentation, volcanism, crustal movements and similar geologic processes. Local catastrophism of intermittent character is considered acceptable today. But a great cataclysmic geologic upheaval such as the world-wide Flood spoken of in the Bible is almost universally repugnant to modern geologists.

Biblical catastrophism, we see, accepts uniformitarianism in regard to physical law. The conservation law has, as can be seen, been in effect since the end of creation and the entrophy law since the imposition of the curse. The essential uniformity even of process rates is accepted and acknowledged for the post-Flood period. The basic geophysical processes of the rotation of the earth and its orbital revolution, which influence most other biological and geological processes, were pronounced as fixed by God.[28]

Creationists suggests that this general uniformity has to be sufficiently elastic to allow for wide statistical variations in rate in all ordinary processes and also for the possibility at least of miraculous intervention by God in either law or process whenever he so wills.

Geologic catastrophism in some form is necessary to explain the phenomena found in the earth's crust, especially the fossils preserved in the rocks, to provide documented evidence for evolution. Billions of plants and animals buried in sediment have been preserved against predatory scavengers, decomposition, etc., even if they had to be buried quickly and permanently. There are countless fossil "graveyards" all over the world.

The fossil record is the only real evidence for large scale evolution, the concept which involves the slow uniform operation of natural processes over great spans of geologic time.

The relative dating of the formations in geologic time, in fact, is itself accomplished on the basis of the state of evolution of the contained fossils; and one can only marvel at the professed ability of the historical geologist to decipher the world's past geologic history.

Catastrophism and not uniformitarianism seem to better interpret most of the earth's past geologic deposits, even though uniformity prevails in the present, say the creationists.

To recognize the necessity of catastrophism as a legitimate geologic principle invites the entrance of all kinds of imaginative catastrophic and quasi-catastrophic theories; whereas the uniformitarian hypothesis has a greater strength and advantage, because it states that geologic interpretations are supposed to have been developed within limits of actual known geologic processes as they exist today.

Catastrophism, though, has no bounds. A catastrophe may be postulated to fit any geologic phenomenon; and yet there would be no way to scientifically prove or disprove the "event" since it is non-reproducible in the scope of scientific observation and experimentation.

There have been many theories of catastrophism, one (in 1950) by Immanuel Velekovski, who wrote books of a modern theory of catastrophism involving a series of encounters of the earth with large comets that later became planets. Also, we have the wobbling axis theory of H. A. Brown.

Among orthodox geologists has come support for such neocatastrophic concepts as shifting poles, drifting continents, asteroid and comet encounters, wide-spread floods, and many others. So we can see from these few examples, catastrophic speculation has become more acceptable in recent years and old style uniformitarianism has encountered some criticism.

The main problem with catastrophist theories is that there is no way to test them empirically. One can imagine all sorts of things but have no way of proving it. There is really no restraint on imagination or speculation when speaking of catastrophism, and this is one reason it has been in such poor repute for over a hundred years.

Creationists agree that catastrophism is, as we have seen, necessary. They feel we do not have to speculate, though, because the Biblical record has provided a clear description of the causes, nature, and results of the actual, true catastrophism. The only world-wide catastrophe mentioned in the Bible, God's word, is the Flood; and it is more than adequate to account for all the earth's physiographic and geological evidences of catastrophism. Although it is true that we cannot verify it experimentally, any more than we can any of the various other theories of catastrophism, we do not need experimental verification.

The creationists suggest that all the fossil records and all ancient plants and animals were formed during a single world-wide catastrophic flood. That flood lasted almost 300 days and produced the entire fossil record as we now know it. According to Dr. Henry Morris, the geologic column does not represent the slow evolution of life over many ages, as the evolution model alleges, but rather the rapid destruction and burial of life in one age, in accordance with the creation model.

The evolutionists suggest that the flaws, fallacies, and contradictions of this model are almost too numerous to recount. Every single evolutionary series which has been documented (and, gaps and all, there are hundreds of such series) would have to have been produced by the sorting action of the flood waters. In other words, when a remarkable series of evolving forms like the series leading to the modern horse, higher plants, etc., are considered, the special creationists must teach that all the various forms and species were in fact created at the same time! Thus, the more one considers the fossil record, the more laughable the notion of "flood geology" becomes.

EVIDENCE FROM SIMILARITIES

Some Structural Similarities

There are other facts about living organisms besides fossils that are used as "evidence" for evolutionary theory. They come from some of the subsciences of embryology, morphology, physiology, biochemistry, ecology, and taxonomy.

Data on similarities of organisms from the above areas of study help support the evolutionary theory. Evolutionary reasoning is based on the following assumption: "the degree of similarity of organisms indicated the degree of supposed relationship of said organisms."

If the animals look alike, they are closely related; if not, they are not. These "evidences" from similarities can be used to demonstrate genetic relationships. Comparative anatomy (deals with animal structure) is thought by evolutionists to show common ancestry of animals. In comparing the forelimbs of a salamander, crocodile, bat, bird, horse, and man, they all have an upper arm bone (humerus) and two lower ones (radius and ulna). These are homologous structures--they are similar in type of structure and origin but not necessarily in function. (For example, the bat's wing is homologous (matching in structure) to a dog's forefoot but not to an insect's wing. This similarity in the two structures does indicate they were derived from a common ancestor!)

If we could see these limb bones on each of these animals, we would see that each of these land vertebrates have "hands," but that the hands have differences which allow each animal to fall into its natural place.

The main function of these appendages are for mobility and obtaining food.

Creationists believe that when God created vertebrates, He used a single blueprint for the body plan but varied the plan so that each "kind" would be perfectly equipped to take its place in the wonderful world He created for them.

If all organisms, as the evolutionists claim, have a common ancestor, then there should be a continuous intergradation between all the different kinds of animals and plants. Instead, though, there are gaps between the kinds, both in the fossil record and today. Only special creation with variation within limits can account for both similarities and differences.[29]

Some Developmental Similarities

Embryology

The recapitulation theory is based on embryology, the study of the young before emerging from the egg. This theory says that the embryo of an organism progresses through stages of embryonic development in which the stages of its evolution can be seen. The basis for the theory is that the embryo begins as one cell (the protozoan stage) and becomes more complex. Some stages were thought to have a resemblance to lower animals in some structures.

One main example was the human heart. It was believed to have passed through a worm, fish, frog, and reptile stage before reaching its final form. It is a fact that in one of its stages the heart of the human embryo was one chamber (as the worm), two chambers (like the fish), three chambers (the frog) and four with a connection of the two sides (reptile). The "reptile stage" is necessary to shunt the blood around the lungs until after the child is born. Because the embryo's oxygen comes from the placenta before birth, there is no need to send a large supply of blood to the lungs.

In recent years, the recapitulation theory has been all but abandoned. Scientists in research have shown these stages as necessary for the embryo's development from one cell to a complex organism. The heart develops as demands on it become greater. Structures in the human embryo which resemble those in animals do not, in many cases, ever have the functions of similar structures in other animals. "Gill Slits" in the human embryo never have a respiratory function as in the fish. "Worm and frog stage kidneys" probably never have excretory functions at any stage. Several stages that would be expected to be found are absent or, as in the early stages of the heart, occur in the "wrong" order. There is no animal that resembles the embryo in its early stages.

Vestiges

Vestigial organs are those structures for which no use has been found. There used to be a list of one hundred or more, but now there are very few because man has discovered the uses of all the others.

Supposedly the appendix, the ear muscles, the third eye lid or nicitating membrane were organs man used at an earlier evolutionary stage. Evolutionists believe man's appendix was once larger and used it, before his food habits changed, as a digestive organ. According to the theory, the appendix is now a small vestige of its previous usefulness. Now we have evidence that man's appendix serves in the body's defense against disease. Ancient man is represented by only bone fossils; so we cannot determine what early man's appendix was like.

The nicitating membrane is not vestigial, but functions in collecting foreign material that gets into the eye and in depositing it in a sticky mass in the eye's corner.

The coccyx is thought by some evolutionists to be a vestigial tail. During embryonic development, it temporarily protrudes. Usually when born, the child's "tail" has long since been enclosed in the tissues. It does not shrink but is surrounded by the developing hips. It serves as a place of attachment for certain muscles. This is an important function because without this bone we could not sit comfortable since important parts of the muscular system are attached to it.

Explaining all apparently "useless organs" as vestiges may lead to some difficult problems. Male mammals possess non-functioning mammary glands. If they are vestigal, then males must have at one time suckled their young.

Once endocrine glands were considered vestigial. Now we know they are important in producing hormones. The thymus recently was discovered to protect the body from diseases.

Creationists believe there are weaknesses in the vestigial organ argument. First, as has been pointed out, just because the function of an organ is not known does not mean that the organ is useless. When it is removed and no damage is caused, it could be because other organs take over its function. Second, why would not using an organ cause its deterioration? This is Lamarckian reasoning and has certainly not been demonstrated. Even if the organ were never used, the formation of the organ by the genes would not be altered. Supposed deteriorations occur because of "loss mutations," but why would these individuals be selected when they were losing the use of an organ that, if not beneficial, was certainly not harmful?

Would any of the previously discussed "vestigial organs" have a harmful effect on the person they belong to? Mutants in some cases tend to have less vigor than normal individuals. So the mutant would really have less chance of survival, not more. The evolutionist feels that the vestigial organs can to a certain extent be used to support evolution toward increasing complexity; because, if the individual has fewer non-functional organs, it may show higher levels of sophistication than those in use.

Biochemical Similarities

Scientists have turned to physiology and biochemistry, in recent years, as bases for investigating evolutionary relationships by arguing that body chemistry gives a more valid basis for conclusions regarding relationships among organisms. DNA (Deoxyribose Nucleic Acid) is found in the nuclei of all living things, and ATP (Adenosine Triphosphate) is involved universally in energy transfers of organisms. Because of this, evolutionists say that all living things had a common origin. Hormones and enzymes in various mammals are so similar that they are used to treat human diseases. Insulin is an example.

Even though DNA is a constitutent of all living organisms, it is very highly organized and complex. All the genetic "information" necessary for the production of each "kind" of animal or plant is found "enclosed" in the DNA. Thus, complex substances permit wide variation in particular traits; however, all these variations in traits are implicitly present in the genetic material at the beginning of every individual organism. There is no way for new traits to come about except by mutations (random changes in the arrangement of the genetic system).

When studying antibodies (blood proteins), some scientists say man is more closely related to the apes. On the basis of blood types, though, he is closer to monkeys. Hemoglobin, a blood protein, is found in most vertebrates, paramecia, plant root nodules, and in the pea family. Very primitive vertebrates and cyclostomes have the same type of hemoglobin found in most invertebrates. Man and dogs can have rabies; man and birds, malaria; man and rats, plague; and man and goats, Malta fever. Can man be equally related to all these creatures? The creationists suggest that body chemistries are involved in any disease and disease similarities defy the evolutionary theory.

Taxonomy

Our current system of taxonomy (the science of naming and classifying) is based on a method proposed by Carolus Linnaeus, a Swedish scientist. He named and classified organisms in order to provide a systematic, orderly way to study and deal with them. This system of noting similarities and grouping organisms under such categories as phylum, class, and order helps to make the world of living things more easily grasped by the student.

Evolutionists suggest another reason for classification: to show evolutionary relationships of the organisms. Because of this approach, taxonomy is many times referred to by evolutionists as systematics, or the study of theoretical kinships.

In systematics, the student looks for the "relationship" of organisms from an evolutionary point of view. At the beginning, Linnaeus believed the various species were created separately; later he believed the created kinds were capable of great variation. As mentioned before, this is the viewpoint held by creationists today (but they do not consider created kinds to be species as we know them today). Linnaeus' system is used today. Creationists believe the various kinds, such as cats and dogs, were created with similarities and differences; whereas, evolutionists believe the characteristics resulted from evolutionary changes.

Creationists feel that evolutionists should stop trying to classify according to presumed evolutionary history and start using structure or biochemistry as the sole basis on which all can agree. Classification could be based on verifiable evidence

EARLY MAN

The Classification and Uniqueness of Man

The origin of man is not solved by history, and there is much disagreement as to whether man came from God's plan and handwork or whether he came up from the brute.

> And God said, Let us make man in our image, after our likeness and let them have dominion over the fish of the sea, and over the fowl of the air, and over the cattle, and over all the earth, and over every creeping thing that creepeth upon the earth. So God created man in his own image, in the image of God created he him; male and female created he them. And God blessed them, and God said unto them, Be fruitful, and multiply, and replenish the earth, and subdue it; and have dominion over the fish of the sea, and over the fowl of the air, and over every living thing that moveth upon the earth. (Genesis 1:26-28)

Though no one knows the age of the human race, we do know man lived prior to the time of written records. So we must turn to that record of this prehistoric era which does remain: skeletons, tools, and weapons found in rock layers.

Animals, as discussed earlier, are classified according to their structure. Although classification should be uniform and standard, it is in fact, subjective and results in differences in grouping by various authorities.

Some structures should be emphasized more than others when grouping living things. For example, vertebrates are placed in the same group, based on a series of bones, the "backbone." If hemoglobin in the blood were the criterion, the earthworm would be included also in this group. The crayfish and grass-hopper are members of the same phylum, because they have an exoskeleton. If they were classified instead, by their way of carrying oxygen, they would be classified in separate phyla because the grasshopper carries oxygen in a system of tubes called the trachea, whereas crayfish carry it in the blood.

We do not need to reject the classification system based on form; however, we should not draw unwarranted conclusion of kinship from it. It should be understood that classification means in essence that animals placed in the same groups have certain physical likenesses and not that similar animals are necessarily related.

Man's classification places him in the phylum *Chordata*, subphylum, *Vertebrata*; class *Mammilia*; order *Primates*; family, *Hominidae*, genus, *Homo*, species *H. Sapiens*. All mammals have hair and the females have milk glands. In the primate order we include monkeys and apes with man. Apes such as gorillas, chimpanzees and orangutans, and man are homonids. Living men are placed in the species Homo Sapiens, and more and more there is a tendency to include certain fossil skeletons in this species. At the time when many skeletons were first found, they were grouped in another species and a separate genus. For example, the so-termed cave men belonged to the genus *Pithecanthropus* or *Sinanthropus*.

There are many likenesses between man and apes. Man is most like apes in physical appearance; but he is by a great margin, more intelligent than apes, who do not manifest much more intelligence than horses or dogs. Man possesses the ability to reason, communicate, plan, or construct, placing a great gulf between himself and the other animals.

In looking at an ape, we see that the shape of his face is much like that of man. He has nails on his fingers instead of claws. The thumb can be placed opposite any of the other fingers, but not as readily as on the human hand. The larynx, though very much like man's, does not enable him to speak. Most evolutionists have claimed these likenesses link man and ape by descent from a common ancestor. Creationists give another alternative to common descent, namely the two have been created on the same general plan to live in like geographical locations. Creationists believe that we have difficulty when claiming physical likeness leads to common ancestry. It is true that the squid has eyes like vertebrates; yet no one believes them to be closely related.

The differences between man and ape are many and significant. The foot of an ape is made and fitted for clasping tree branches rather than walking, and man uses his hand for grasping. They vary in relation to the pelvic girdle: the ape's anterior ilium bone is a straight slab in contrast to man's basin-like pelvic girdle in which many internal organs are found. In contrast to man, apes usually walk on all fours, and their arms are as long as their legs. Hair is evenly distributed over the body.

The ape brain is smaller than man's, yet the ratio of brain size to body size is of more significance than the actual weight or capacity of the brain; although neither is an indication of relative intelligence. The ape has a more prominent brow ridge than man. The ape's canine teeth are specialized and sharp and prominent and long--so long that if the mouth is closed, they fit beside each other and prevent sidewise motion in chewing. Man's canines are no longer than his incisors.

Not only physical differences set man apart from other animals, but also there is greater magnitude in differences in their behaviors.

An ape will not put wood in a fire to warm himself, even if he is freezing. He may use a stick or stone as a tool but does not see any future use for the tool. If attacked as a herd, apes do not organize in resistance; and they do not carry wounded companions with them when retreating.

One important characteristic of man is that he uses language. Animals utter cries but cannot use words as symbols.[30] Man could never have mastered his environment to this extent without the tool of language. Every tribe has its language, and its grammar is complex.

The Biblical record is precise in regard to man's uniqueness in the world of living things. Man was created in God's "image and likeness." God made man king of the earth with dominion over all the animals. God created them male and female. He is unique in life because he possesses self-consciousness: he is aware of his own existence and identity. No animal has ever had the thought, "I am an animal." If he did, he would cease to be one; because, as far as we can tell, self-consciousness is a unique possession of man on the earth.

Only men are capable of learning from mistakes by precious generations. Only he can "experience" history in this way. In the animal kingdom there is no history: one generation learns nothing from the previous generation.

Only man can reckon moral discernment--and thereby be held guilty. No animal has a conscience or moral nature. Only man can worship God, can recognize the existence of his creator and his own responsibility to God. Never has an animal said in his heart, "Thank you God." Animals do not live beyond the grave--only man in his uniqueness. God has made clear in his Word that the spiritual, moral, mental and social capacities are certainly unique and could never be explained by origin from the animal kingdom.[31]

Mortimer Adler, editor of Encyclopedia Britanica, Great Books of the Western World, reports a change in his thinking about evolution in The Difference of Man and the Difference it Makes. He has concluded that you cannot explain man in terms of physical evolution from the animal. He offers no alternative, however; but he does admit evolution has failed to come to grips with this fact.

Creationists say that God clearly reveals in his Word that our supernatural origin and God-created dignity and responsibility to Him spiritually provide the foundation on which God's complete revelation in scripture is constructed.

The Search and Discovery of Early Man

I would now like to touch on the search and discovery of "Early" man. When a human skeleton is found, the first question probably asked is how old is it? It can be answered indefinitely and in many ways. We might think that relative ages could be estimated by the depth of burial. This would be true only if the body were buried by natural agencies and not by people. But depth is often not considered because most are found near the surface of the earth or in caves.

Some scientists will judge the age from the form of the skeleton--if well-formed, it must be recent; if poorly-formed, ancient. This kind of reasoning is based on a belief of evolution. If the doctrine of evolution were proven, beyond a shadow of doubt the reasoning would be valid, but because it is only a theory, the conclusion is only tentative. It may also be true that the deformation of bones may be due to desease as in the Neanderthal people.

There have been many pictures of the "brutish" cave man, almost naked, thick legs, bent knees, stooped neck and prominent brow ridges. Skeletons later found were more complete; and it can now be concluded that Neanderthal man probably stood erect, had strong bones and a large brain--possibly larger than modern man--1450 cc--man's today is no larger.

He made useful tools and weapons, used fire and buried the dead with the things that he thought he would need. One writer said, modern people who consider the tools of Neanderthal man crude could not make the same types themselves.

A Spanish nobleman, M. Santuola, was digging on a cave floor of Altamira in northern Spain. His little daughter was with him and both had a light. The little girl, restless, wandered further into the cave. She called out Toros! Toros! Her father hurried to the rescue and found her looking up at the ceiling. There were paintings of bulls, bison, horses, and boars in color and with marked accuracy.

At first people did not believe that people who lived before history began could create such paintings. It is now believed that they are the work of Cro-Magnon man. He also made stone tools and was a carver of ivory. Cro-Magnon man was tall and well-formed in all parts of his skeleton, and his cranial capacity was 1660 cc--larger than that of present-day man.

If man evolved from animals, we might think such an excellent type lived in recent times. It is hard to say when Cro-Magnon man lived, but his bones have been found mingled with those of Neanderthal man in a cave on Mt. Carmel. So we can deduce that some of the two types lived simultaneously. In France, in 1948, evidence was collected, that some investigators accept, that modern types of men lived before Neanderthal. Germaine Henry-Martin, an experienced archaeologist, found parts of two skulls of a modern type deeper beneath the earth's surface than tools made by

Neanderthal man. A layer of secondary limestone was formed between them after the modern type men died but before Neanderthal men lived.

In 1959, a much fragmented skull was found in a 300-foot-deep gorge in Kenya, Africa. The pieces were taken out of the rock and put together, but the lower jaw was not found. They named it Zinjanthropus. Its teeth were like modern man's in size, shape, and arrangement. This Zinjanthropus is similar to fossils found in southern Africa that some investigators thought were "missing links" between man and an ape-like ancestor. There are arguments against this idea because Zinjanthropus lived at the same times as other early fossil men. One unusual feature about this creature is a crest projecting from the upper part of the skull which is not as prominent as the gorilla's.

A discovery found in 1964 by Louis Leakey further confused the issue of African fossil men. Below the rock layers where Zinjanthropus was found, Leakey found the skull and bones of Homo habilis (able man). The skull, hands and feet are quite similar to modern man's. Stone tools found nearby were believed to have been made by Homo habilis, although they may have been made by Zinjanthropus. Homo habilis is about 4 feet. Fortunately, several other habilis skeletons were found in the gorge 300 feet deep in solid rock. These skeletons are some of the oldest human fossils, but not the simplest.

It has been said that skeletons of modern types have been found as deeply buried or deeper than peculiar types of skeletons, or that modern types were found deeper than the thought-to-be more primitive ones. People who believe evolution occurred believe that the peculiar types are ancestors to modern men.

Even on the basis of the few fossils found, it is true that some unusual types did exist. There are several possible explanations. It is possible that some "man-like-apes" were created and lived contemporaneously with man and apes. Some isolated groups of men may have degenerated. Other unusual characteristics may have resulted from a disease or poor living conditions, for the skeletal system is susceptible to change from the latter causes.

From the genetic studies we can see that a typical mutation usually lacks vigor and has less of a chance in the struggle for existence. The vast majority of mutations have been found to be detrimental to the continued existance of the individual or species.

With these facts as a basis, some creationists believe that most fossil men were simply men of slightly different varieties from those of today.

FOOTNOTES

[25]"Moore, op, cit., p. 425."

[26]Henry M. Morris, Biblical Cosmology and Modern Science (New Jersey: Craig Press, 1975), p. 25

[27]Ibid.

[28]Ibid., p. 26.

[29]"Moore, op. cit., p. 432."

[30]There have been a number of research studies in recent years--some of which are still in progress--to determine the actual communication capabilities of such animals as apes and dolphins. Results of each are subject to interpretation and controversey: for the present, at least, man's communicative uniqueness has not been indisputably challenged.

[31]John C. Whitcomb, Creation Counseling--The Origin of Man, Vol. 3, Tape 2. (California: Creation Life Publishers, 1973).

CHAPTER IV

GENETICS

DEFINITION

Genetics is the branch of science that considers how traits are passed from generation to generation.

HEREDITY AND ENVIRONMENT

By applying genetic principles, man has improved plants and animals and human welfare has advanced. Teh quality of fruits and vegetables has improved through the development of plant strains that are more resistant to disease. Genetics has also brought about a more rapid advance in the production of desirable types of animals. At special genetic clinics people have been advised of problems related to inheritable family traits.

Heredity and environment determine how an individual will develop. Traits for eye color are determined totally by heredity and not by environment. Whether or not one will develop diabetes, depends on both heredity and environment. The size of a child depends on both heredity and environment including, of course, food intake or proper diet.

MENDEL AND HIS LAWS

The basic laws of heredity were discovered by an Austrian monk named Gregor Mendel, who taught high school classes in Brunn, Austria. He did experiments with peas in his garden at the monastery, crossing peas that differed in certain characteristics and observing how these characteristics were inherited in subsequent generations. Two reasons he succeeded where others failed in discovering the principles involved in heredity were:

1. The peas he crossed were nearly alike, but different in one or two simple yet obvious characteristics.

2. He kept a record concerning the ratios of the various kinds of offspring.

In the generation after the cross, the individuals were all alike. In the second generation, most plants have the same characteristics as in the first generation. The ratios show that in generation two there are about three times as many individuals like the first generation as there were of the other type. The ratio is 3:1.

Although people did not know of chromosomes at this time, Mendel decided that hereditary factors had to be in pairs because a characteristic that did not show in the first generation reappeared in subsequent generations. He concluded from this that the factor for the characteristic that did not show up in the first generation was still present. The one that was seen in the first generation he called "dominant" and the other he called "recessive."

Each generation of plants is expected to have the same total number of genes, or hereditary factors; and so the reproductive cells, or gametes, have one of each of these genes instead of a pair. The zygote, or fertilized egg, has a pair of each type of gene--one member of the pair from each parent (allele).

We can use as an example for inheritance the color trait of peas. The color of the peas, green or yellow, is evident upon the appearance of the cotyledon (the first leaves of the new plant). The gene for yellow is dominant; so "Y" will designate this color and "y" will designate the recessive green trait. They (genes) are found in pairs; so the yellow parent is YY, and the green one, yy. When either YY or yy are used, they are known as homozygous, which means their gametes are all alike. The hybrid Yy is called heterozygous, and it produces two kinds of gametes, ½ Y and ½ y. The following diagram shows what happens when two hybrid plants reproduce:

Figure 6

	Y	y
Y	YY	Yy
y	yY	yy

¼ will have YY yellow color

½ will have Yy

¼ will have yy green color

There will be a 3-1 ratio, 3/4 yellow, one pure and one hybrid, and ¼ green yy.

The same results are obtained when other characteristics are used. For example, when pure parent types of round R and wrinkled r seeds are used, the first generation is all round and the second generation of hybrid crosses gives the 3:1 ratio--3/4 round, 1/4 wrinkled.

The phenotype is an expression in words of the characteristic aspect of an observable trait--a yellow pea has a phenotype of yellow. The genotype represents in letters all the hereditary factors for a trait (i.e., all the genes carried, whether or not they are evident in a given generation). A phenotype of a yellow pea may be pure yellow YY, having two genes for yellow, or hybrid Yy, with one gene for yellow and one green. Yet all are yellow because yellow is dominant over green.

Alleles are genes that control different expressions of the same trait (Y and y) and occupy the same location on the same kind of (paired) chromosomes.

After a cross of pure types, the first generation is produced and is known as the first filial generation or F_1; another generation is produced by inbreeding the F_1, and it is called the F_2, or second filial generation. The F_3 and F_4 are not discussed in genetics, because with random inbreeding all generations after the F_2 will be the same as the F_2.

There are two main laws derived from Mendel's work. The first of Mendel's laws is the law of Segregation. It deals with genes of the same pair that are alleles. It states that in the formation of gametes, genes that are alleles and segregate; that is, genes of the same pair go into different gametes not into the same gamete. An example would be if a plant has the genotype Yy, the Y and y will segregate and go into different gametes. The result will be Y gametes and y gametes. There would be no Yy gametes.

Mendel's second law is the law of Independent Assortment. It deals with genes of different pairs, those which are not alleles. It states that in formation of gametes, genes that are not alleles assort at random. An example of this would be: A plant with a genotype of RrYy the gametes will be RY Ry rY and ry in equal numbers. There will be no gametes of the type RR, rr, YY or yy.

Mendel's second law does not hold true if two pairs of genes lie on the same chromosome. If non alleles are on the same chromosome they cannot assort at random. This is evident when visualizing meioses. Most biologists include Dominance and Unit characters as additional laws.

LACK OF DOMINANCE

In Mendel's peas there was always one allele dominant over the other in each pair, but sometimes dominance is lacking in a pair of alleles. The hybrid is a blend between the homozygous parent types. For instance, in some species of red crossed with white flowers neither is dominant and the hybrids are pink. In cattle of the short-horn type, when red and white are crossed, the F_1 is roan having a blend of red and white hairs.

Because of a lack of dominance, three phenotypes are expressed (color characteristics)--red, roan and white 1:2:1.

MUTATIONS

Mutations are changes in the gene itself and are expressed in the phenotype. Mutations have been produced experimentally by the use of high energy radiation, heat, or certain chemicals. They (mutations) can also occur in nature and are caused by radiation.

Most mutations are harmful; however, a few may be good in an unusual environment. Most geneticists, however, accept the fact that less than one mutation in a hundred is considered good in any environment and some say one in one thousand is more accurate. Some geneticists say the number of mutations that might be considered advantageous to the individual is so small, if indeed there are any at all, that mutations may be deleterious. Indeed, many mutations result in death.

A mutation that produces an advantageous change in an unusual environment usually can be injurious physiologically; and these effects will weaken the individual. Consequently, even the rarely-occuring "good" mutations may be actually detrimental to the species.

SELECTIVE BREEDING

Genetics can be applied by man in selective breeding. Archaeological evidence indicates that from the beginning man used domesticated plants and animals and that he has always attempted to improve the quality of such organisms. Selective breeding was the first method he used, choosing the best and finest plants and animals as breeding stock and selecting them for specific desirable characteristics.

An example of selective breeding of animals occurs when cattle are bred for higher milk or butterfat production, or when beef cattle are bred for better beef. Plants, too, have been improved by selection. The American Indian knew something about selective breeding when he bred popcorn, field corn, and sweet corn. Even though these breeds hybridize freely, he knew how to maintain pure lines although he had no knowledge of genes or chromosomes.

Today's geneticists still use selective breeding but combine it with Mendel's Laws and other useful knowledge about heredity. Wheat, for example, can be developed to produce good flour and ripen early, so it will not be killed by early frost, and also to be rust resistant. This is a good example of the three characteristics geneticists try to develop:

1. Desirable quality for a specific purpose.
2. Disease resistance.
3. Ability to grow in (a) specific environment(s).

Scientists have genetically produced breeds of hornless cattle which cannot gore each other or the men who handle them. To develop such characteristics, man used mutations and other kinds of genetic variation. The first hornless cow was produced by a mutation. Mutants usually are recessive and breed true. However, the characteristic of hornlessness proved to be dominant and the horned characteristic recessive.

COMBINING DESIRABLE CHARACTERISTICS

Geneticists can combine desirable characteristics. Many characteristics are controlled by numerous pairs of genes, and from man's point of view, these are usually the important and useful characteristics.

Mendel made crosses of yellow round seeds with green wrinkled ones. In the first filial generation he had all yellow and round seeds which showed that these were the dominant characteristics. In the second filial generation he obtained 315 yellow round, 108 round green, 101 wrinkled yellow and 32 wrinkled green. This reduces to a ratio of 9:3 : 3:1.

Considering each characteristic separately, they segregate and recombine the same as in simpler crosses, for there are three times as many round as wrinkled and three times as many yellow as green.

This is an example of the formation of new varieties by segregation and recombination which has been accomplished by breeders many times since. In this experiment, Mendel made two varieties which he did not have in the beginning--yellow wrinkled and green round. These varieties illustrate segregation and recombination of gene material.

Mendel started with yellow in the same plant, with round green in the same plant with the wrinkled, yet a portion of these genes separated--an illustration of segregation. They then lined up with different (new) genes: green with round, and yellow with wrinkled--illustrating recombination. Numerous crosses have been made to bring together two desirable characteristics from distant places into the same variety of plants or animals. An example of this is the experiment by W. B. Keeney, who made the Burpee stringless Green Pod Bean by bringing together the characteristics of thick tender pods with the absence of strings. It was not really new, but new only in that it combined already known characteristics.

The creationists feel that segregation and recombination do not give evidence for the supposed process of evolution, the doctrine that all living things including man developed by natural changes from lifeless matter. The process of evolution would require new and improved traits. Segregation and recombination merely provide a reserve pool of latent genes that may help a species to survive when the environment changes.

MENDEL'S WORK NOT ACCEPTED

After Mendel had worked for six years, he wrote a paper producing the results and principles he deduced from them. Today the paper is considered to be scientifically clear and very accurate. However, when in 1865 he read it before the National History Society of Brunn, Austria, there was no discussion and it went unappreciated. There were able scientists there, but their minds were on other matters. In the minutes of their meeting, it was recorded that later that evening they discussed Charles Darwin's book written a few years before entitled The Origin of the Species. Mendel's paper was not appreciated until 1900, thirty-five years later when it was rediscovered by three geneticists whose separate investigations were similar in results. After 1900 developments in genetics occurred rapidly as they do today.

JOHANSEN AND SELECTION

For hundreds of years, farmers and livestock breeders have attempted to improve plants and animals by selection. They have chosen the best individuals from their gardens, fields, and herds to be parents of the next generation. Many sold the best wheat and young cattle to bring the most money on the market, but other farmers saved the best for seed and usually profited by so doing. If a certain type in a population is chosen to become the sole parents of the next generation, that generation will have a different average from what would have appeared if the parental stock had not been selected. This is known as effective selection. Not all selection, however, is effective.

Selection is, many times, most practical for traits governed by multiple genes. They occur in pairs--in chromosome pairs just like the simpler ones--but in more than one pair of genes, each pair controlling a certain part of the weight.

Wilhelm Johansen (1855-1927), a Danish botanist, raised the Princess variety of beans. He selected them until he had strains which he called "pure."

In this species a pure line consists of the progeny of a single bean. One bean plant within a pure line has genes just like those of any other plant in that line; however, the plants may be of different sizes due to the soil differences. The size of the beans varied also because the ones in the middle of the pod had more room than those at the ends and therefore grew larger.

Johansen chose the largest beans in a pure line, carefully weighed them and planted them in a little plot of ground. Next, he chose the smallest beans of that same pure line and planted them in another plot. He harvested the plots separately and then weighed the beans. The results were surprising, for his selection, though properly done, was done ineffectively. Rather than big beans producing big beans, the beans produced in the two plots were of the same average size.

These unexpected results caused a stir among geneticists. In 1910-1920 they tried similar tests on other species. In selecting fewer bristles on the fruit fly thorax, there was a reduction for twenty generations, but after that the number of bristles remained constant. After a number of tests in various species, all geneticists agreed with Johansen that a condition is reached in which the genes affecting a trait are alike, and after that time it makes no difference which individuals are chosen: the progeny are the same. This situation seemed unusual and strange to geneticists because in the nineteenth century they were taught that genes are changing continually in all directions. Scientists realize change is a rare occurrence, and it is called a mutation.

This principle is true in other species as mentioned, one example of which is the improvement in the quality of sweetness of the sugar beet. Results were the same as with the beans and the fruit fly.

It follows, too, that hybrid corn is an example of gene stability. The first step in its production is to self-pollinate itself for at least six generations. In the corn progeny many various abnormal and undesirable characteristics appear, which are discarded. Some are discarded because they fail to produce seed. The vigor even of the best of the inbred strains is reduced after six generations of self-pollination. There is no further reduction in vigor and many very uniform strains have developed. Within any one of these strains, selection is equally ineffective, as in Johansen's beans, because the genes are homogeneous and homozygous.

To produce hybrid seed corn, two of the inbred strains are planted in rows alternating with each other and the tassels of one strain are removed to make them cross with the other strain. When these seeds grow into mature plants the original vigor is restored and increased because undesirable genes have been discarded.

Farmers who breed corn do not hybridize all their inbred corn in any year, because they must keep part of each inbred strain pure for hybridizing in future years. Thus there have been remarkable changes made in Zea Mays, but no one thinks it is changing into another species.

"MUTATIONS--PROFIT OR LOSS"

When scientists finally agreed that Johansen was accurate in his conception of gene stability, they fell back on mutation as the reason for evolution: changes due to the environment, the so-called acquired characteristics, are a factor in evolution. Creationists suggest, similarly, that segregation and recombination of characteristics could not ever build up complex organisms from simple materials because they do not supply any new characteristics; instead, they separate characteristics as often as they combine them. For proponents of evolution, then, mutation is the only mechanism left to explain new animal and plant forms.

An example of a situation in which a mutation occurs and a mutant characteristic is produced is the case of fruit flies (Drosophila melanogaster). In an experiment they are kept in half-pint milk bottles with a cotton stopper to let air in and one-half inch of mashed banana or similar food in the bottom. Eggs that are laid on the food produce larva which live in the food and, when fully grown, climb up the side of the bottle and change to the pupa or resting stage. Next, in order to be examined, they are moved to an empty bottle and kept inactive by placing a small amount of ether on cotton. Later they again become active.

In one of these cultures a fly came out of the pupa stage with white eyes instead of the usual red ones. This change was a mutation, and the characteristic of white eyes was a mutant. This fly, a male, was then placed in a bottle with a red-eyed female (there were no white-eyed females). The offspring produced were red-eyed, which proved red was dominant over white. In the third filial generation part red-eyed and part white-eyed offspring were found. (In the F_2 the females were all red because eye color is sex linked.) As a result of mating the white with white of the third filial generation, the trait was suppressed. Mating of this type has preserved and maintained this mutant strain until the present time. This has only been accomplished in laboratories for the most part.

A mutation occurs in a single animal or plant; it takes place suddenly and affects one characteristic but reduces the general vigor; and is inherited in the same way as other characteristics. So far as scientists have been able to learn, the production of mutations is very sporadic. There has never been a way either in the laboratory or nature to produce a certain mutation according to pre-arranged specifications. Scientists have discovered such things as x-ray, mustard gas, radium rays, and ultraviolet light do increase the rate of this change; however, the kinds of mutants produced are not different from those that happen for unknown reasons.

A valid scientific method of learning the nature of mutants is to enumerate some that have been observed. There are some spoken of in the literature which have only been supposed and have never been observed. These are different from the ones that have been seen, examples being--increased strength of wing muscles, greater speed of rabbits, etc. Mutants found in the fruit fly include changes in the wings: small miniature, notched, nicked, redimentary, vestigial, and no wing. There are many other examples of plant, animal, and human mutants. Geneticists agree that most of all mutations are lethal. Indeed, the homozygous condition (paired genes for a particular gene identity) could cause death.

Others have claimed that any other mutation confers an advantage upon the plant or animal that has it, except when an environmental change occurs; and usually examples of this type are hypothetical. Never has anyone observed mutations occurring that would change one class of animal into a more complex type of organism; for example, the beginning of a milk gland upon a reptile's breast, changing it to a mammal; or a feather starting in place of a scale, transforming it to a bird. The changes that have been observed involve a loss in some physical trait.

Creationists suggest that, considering the discoveries of the twentieth century, if one still believes that mutations have changed an amoeba to man, the reason would have to be that he has faith that, in times past, mutations took place that were different than those that now occur. Intellectually, he prefers a system of belief which violates a belief in the uniformity of nature.

There is sufficient evidence that Mendel's laws not only apply to plants and animals, but also to human beings. However, it is impossible to study human genetics as we study other organisms. One reason is that we do not control the mating of people, and few people keep family records. Geneticists study twins and families to learn about human heredity.

Geneticists have made studies of traits such as color of eyes, skin, hair and body size. But, obviously, these are not as important as studies of diseases, intelligence and blood types.

THE INHERITANCE OF SEX

In regard to inheritance of sex, both in man and Drosophilia, there is one pair of chromosomes in the male that are not the same size. The small y chromosome contains almost no genes. The large Y chromosome contains genes that have no paired gene. The female has two X chromosomes which makes the genes paired as in the regular chromosomes. This simple diagram illustrates how sex is determined in man using X and Y to represent chromosomes not genes:

Figure 7

Male XY
Female XX

Eggs	X chrom
Sperm X chrom	XX(f)
Y chrom	XY(m)

Usually the average of the offspring are of each sex; but, actually, there are a few more boys than girls born.

"SEX LINKED TRAITS"

In sex-linked traits a mother may be carrying a gene factor from color-blindness from father. The gene comes from the X chromosome and is recessive. However, from her mother she received the gene for nowmal vision. The color-blind gene will be noted as X^c and the normal X^C. The chart and interpretation of the chart will show how color-blindness is passed on:

Figure 8

Mother X^CX^c
Father X^CY

Eggs		X^C	X^c
Sperm	X^C	X^CX^C	X^CX^c
	Y	X^CY	X^cY

The four possible combinations include:

X^CX^C-------------------Normal Girl
X^CX^c-------------------Normal Girl carrying recessive gene
X^CY-------------------Normal Boy
X^cY-------------------Color-blind boy who can pass color-blind gene to daughter but not to son.

Each child born may have been one of the previous four phenotypes. Color-blindness is expressed in males, because no homologous region in the Y chromosome exists to suppress it. In other words, there is no region carrying a normal allele. In females, the X chromosome which carries the normal gene suppresses the recessive allele that is defective. If a girl's mother is a carrier, and the father is color-blind, the girl may receive two genes for color-blindness and so be color-blind.

Figure 9

Eggs		X^C	X^c
Sperm	X^c	X^CX^c	X^cX^c Girl
	Y	X^CY	X^cY Boy

One other sex-linked trait is hemophilia, or bleeder's disease. In this disease the blood does not clot normally because of a missing substance. Children who have the disease bruise easily and can die from unstoppable bleeding. Doctors have progressed some in treating but not curing this disease. Many boys do not survive infancy or early childhood; but if they reach adulthood, many times they are crippled from bleeding joints. They usually do not pass on this very rare gene; because, even if they do survive, they do not marry and become fathers because of their poor health.

This gene, like other defective ones, may occur by mutation. Queen Victoria, it was believed, received the faulty gene in this manner. Because royal families married close relatives, the trait appeared frequently among her descendents, and this had important effects in European History.

INHERITANCE OF INTELLIGENCE

Intelligence is also considered hereditary. A reliable definition of intelligence cannot be found. Many factors and pairs of genes are involved, it is believed. In pursuit of understanding of such matters, geneticists have studied several families and twins. They have found that identical twins (which develop from a single fertilized egg) have practically the same level of intelligence; whereas, the intelligence of other children in the same family may vary more. Usually children resemble their parents and brothers and sisters more than other people. Mental capacities vary in a family as does hair color or size. Just as physical growth is modified by environment, so also mental development is affected by environmental factors. Thus parents should provide the best environment to help their children reach their greatest potential.

Some kinds of feeblemindedness have been found to be hereditary. Mongoloid imbeciles, who have a form of feeblemindedness, have been shown to have an extra chromosome. Phenylpyruvic idiocy is a hereditary condition wherein an enzyme is lacking--a condition caused by single recessive gene. Some types of mental illness are believed to be hereditary, but I believe they are environmentally induced: that an imbalance occurs in reaction to an emotional or mental stress.

NATURAL SELECTION, VARIATION AND RECOMBINATION

When Charles Darwin first published his theory of the origin of species by natural selection, he believed that the continual small variations between individuals of a species that are observed in nature, confer differing degrees of advantages or disadvantages in the struggle for existence. The ones with significant advantages would be favored by natural selection, would survive longer and thus transmit the "good" characteristics hereditarily to their descendants. Eventually and gradually new and higher types of organisms would come about.

Normal variations were later found to be subject to the rigid Mendelian laws of inheritance representing nothing really novel, but only characteristics already latent within the genetic system. Molecular Biology of today, with its insight into the genetic code implanted in the DNA system, has further confirmed that normal variations operate only within the range specified by the DNA for the particular type of organism. Further, no novel characteristics producing higher degrees of order and complexity are observed. Variation is not vertical but horizontal.

However, this type of normal variation is offered still as evidence of present-day evolution.

One example of this is the peppered moth of England "evolving" from a dominant light color to a dominant dark color as a result of tree trunks becoming darker due to pollutants (scot) from factory chimneys from the industrial revolution. From 1845 to 1895 the black strain increased in frequency 90-95% as shown by British scientists, A. Fisher and E. V. Ford.

Kettlewell did experiments by releasing marked moths in the industrial area of Birmingham where 40% of the dark moths were recovered at night but only 19% of the light-colored ones were recovered. However, in the Dorset countryside, 6% of the dark moths were recovered, compared to 12.5% of light ones. The tree trunks near Dorset were light gray.

An aspect of particular interest about these investigations that is rarely mentioned is the survival of more than 1% of the normal strain (white) even under very adverse environmental conditions.

Despite natural selection by predators, the light moths still survived. The reason for this phenomenon when the reproductive rate of the white moths was observed to be higher than that of the melanic (black) strain. Even after 166 years of adverse environmental conditions, the "normal" type still persists. The cities of Birmingham and Manchester now use less coal; so the trees are returning to their normal gray color. The gray moths are increasing even in industrial areas due to higher reproductive rates.

Natural selection is a conservative force that operates in preventing kinds from becoming extinct when the environment changes.

The (peppered moths) experiment beautifully demonstrates natural selection--or survival of the fittest--in action, but perhaps does not show evolution in progress. The populations may alter in their content of light, intermediate, or dark forms (Biston Betularia). Instead of explaining evolution the way Darwin thought it did, perhaps the variation phenomenon in natural selection is a perfect example of the creationist's principle of conservation in operation.

A fundamental prediction from the creation model is that since the Creator had a purpose for each kind of organism created, he would institute a system which would not only assure its genetic integrity, but would also enable it to survive in nature. The genetic system would maintain its identity as a specific kind, while, at the same time, adjust its characteristics (within limits) to changes in the environment. If this were not true, then any slight change in its food supply, for instance, could cause it to become extinct.

Natural selection acts as a sort of sieve, allowing only the varients which fit the environment to go through. Those who do not fit do not pass through and are discarded by the sieve. There are only certain genetic potentialities in these variants found in the DNA structure for its particular kind which natural selection can act upon. Natural selection cannot generate anything new itself. Reshuffling or recombination of already present characters in the germ cell does not create anything new in the evolutionary sense. Yet this phenomenon of recombination in natural selection is regarded as an important aspect of the evolutionary model. Recombination of material for natural selection is by far the most important source of genetic variation.

The creationists suggest that, even if variation could produce something novel for natural selection to act upon, the novelty would be eliminated quickly because it would not be fully developed. For example, a wing on a previously earth-bound animal would be useless and harmful in its non-developed state and so eliminated. Yet evolutionists believe if their model is correct, wings have "evolved" four times: in flying insects, reptiles, birds, and rats.

Similarly, the eye has evolved independently three times: in the squid, vertebrates and anthropods. You may recall that Darwin said to think how the eye could have been produced by natural selection made him ill.

Natural selection acting upon the variational potential that is designed into the genetic code for each organism is a powerful device that permits horizontal variation or radiation, enabling the organism to adapt to the environment and so to survive. Natural selection serves also in generating a vertical variation leading to the development of higher and more complex kinds of organisms. In fact, it acts to show such vertical variation since insipient novelties would be at best expressing developed and functional qualities.

GENETIC MUTATIONS

Creationists feel that neither ordinary variations nor recombination of existing characters can account for "upward" evolution; instead, some extraordinary mechanism must be found for this purpose. In the modern synthetic theory of evolution (or new-Darwinism), the mechanism universally adopted for this purpose is that of mutation.

It is assumed that a mutation is a real structural change in a gene in which something novel is produced, not just a reworking of something already there. Somehow the linkages in a segment of the DNA molecule are changed causing different "information" to be conveyed via the genetic code in the forming of the structure of the descendant.

It must not be forgotten that mutation is the ultimate source of all genetic variation found in natural populations and the only new material available for natural selection to work on.

The mutation phenomenon is a very important component of the evolution model. This model must postulate some mechanism to produce the required upward progress in complexity which is characteristic of the evolutionary model. The mechanism is mutation.

So the evolution model would seem to support the idea that mutations must be primarily beneficial, generating a "vertical" change upward toward higher degrees of order. Any change would have to be positively helpful in the environment if it is to be preserved by natural selection and contribute to evolutionary progress.

Creationists, however, predict that if real mutations do exist, causing "vertical" changes in complexities and order of the kind, they would not be beneficial but harmful.

In thinking of the two models we can consider some actual experimental facts that are relative to mutations:

1. Mutations are random, not directed.

It remains true to say that we know of no way other than random mutation by which new hereditary variation comes into being, nor any process other than natural selection by which the hereditary constitution of a population changes from one generation to the next.

2. Mutations are rare, not common.

It is probably fair to estimate the frequency of a majority of mutations in higher organisms between one in ten thousand and one in a million per gene, per generation.

3. Good mutations are very, very rare.

H. J. Muehler said in regard to experimental observation of mutations:

> Creationists continue to espouse the idea that mutations are found to be of a random nature, so far as their utility is concerned. Accordingly, the great majority of mutations, certainly well over 99 per cent, are harmful in some way, as it is expected in the effects of accidental occurrences.

Julian Huxley, also responsible for the modern view of evolution (neo-Darwinism, which states that evolution proceeds by accumulation of small mutations preserved by natural selection), is concerned about the frequency of beneficial mutations when he suggests:

> A proportion of favorable mutations of one in a thousand, does not sound like much but is probably generous since so many mutations are lethal, preventing the organism living at all, and the great majority of the rest throw the machinery out of gear.

A truly beneficial mutation, one that is known to be a mutation, not just an already-present latent characteristic in the genetic material which beforehand lacked an opportunity for expression, and one also which is permanently beneficial in the natural environment has yet to be documented, according to the creationists.

They also suggest that mutations are more than just changes in heredity; they also offer viability, and, to the best of our knowledge, invariable affect it adversely. Does not this fact show that mutations are really assults on the organism's central being, its basic capacity to be a living thing?

4. The net effect of all mutations is harmful.

Even if mutations are not harmful enough to cause individuals to be eliminated completely by natural selection, the result is a gradual lowering of the viability of the population.

The large majority of mutations, however, are harmful--even lethal--to the individual in whom they are expressed. Such mutations can be regarded as introducing a "load" or genetic burden, into the pool. The term "genetic load" was first introduced by the late H. J. Mueller, who recognized that the rate of mutations is increased by numerous agents man has introduced into his environment, notably ionizing radiation and mutagenic chemicals. Creationists suggest that the net effect of mutations, rather than being beneficial, is harmful to the supposed progress of evolution. This is made very clear by the fact that evolutionists are trying to be rid of mutation radiations and are trying to remove them from the environment.

However, the most important actions that need to be taken are in the area of minimizing the addition of mutagens to those already present in the environment. Any increase in the mutation load is harmful--if not immediately, then certainly to future generations.

5. Mutations affect and are affected by many genes.

The mutation concept has become more complex. Now, instead of a certain characteristic being controlled by a specific gene, it appears that each gene affects more than one characteristic and every characteristic is controlled by many genes.

Moreover, despite the fact that a mutation is a discrete discontinuous effect of the cellular chromosome, or gene level, its effects are modified by interactions in the whole genetic system of an individual.

This universal interaction has been described in deliberately exaggerated form, according to the creationists: "Every character of an organism is affected by all genes and every gene affects all characters." It is this interaction that accounts for the closely knit, functional integration of the genotype as a whole.

So it would seem that if a mutation is very highly likely to be deleterious, a changed characteristic requires the effect of several genes and many concurrent mutations. Therefore, the probability of harmful effects is greatly multiplied. Does the probability of "good" mutations in all the genes which have to do with or control a certain character reduce then, to practically zero?

DNA AND RNA

Chromosome replication is actually DNA replication whereby the DNA makes an exact copy of itself.

DNA is the "machine" in the chromosome of the cell that duplicates the chromosome, including all the hereditary features to be passed on to the daughter cell whenever cell division occurs. DNA controls the color of the eyes, body shape, and millions of other facts about a person which make him different from a dog or a horse or any other person. The DNA in every cell of the dog or horse gives them their special characteristics: that cause them to be what they are, not different sorts of beings.

DNA duplicates its information and passes it on to its daughter cells. DNA is a complex molecule built in the form of a double spiral, or helix. These two strands of the helix are long chains of alternating sugars and phosphate groups; this double spiral is joined by four ends of connecting rods ATG and C (adenine, thymine, guanine, and cytosine), and they appear in various sequences. There are so many possible sequences that it has been estimated that there are at least four or five billion different kinds of DNA molecule combinations in man's forty-six chromosomes, each of which controls one of his features. This is comparable to the twenty-six letters of the alphabet which combine into a dictionary full of words.

The code contained in DNA directs the various cellular processes. First a gene (a segment of a DNA molecule) directs the formation of a molecule of "messenger RNA", thereby transferring to it the gene's instructions for synthesizing a specific protein. As soon as it is formed, the messenger RNA diffuses out of the nucleus and goes to one of the many ribosomes located in the cytoplasm of the cell.[32]

Ribosomes are the cell's tiny protein factories. They consist of protein and RNA (ribosomal RNA). Next a transfer RNA molecule found in the cytoplasm attaches itself to a molecule of a specific amino acid and transfers it to a ribosome where it fits into its proper position as indicated by the messenger RNA. Then more transfer RNA molecules, one after another, bring more amino acids to the ribosome and fit them into their proper positions. The result is a chain of amino acids joined to one another in a definite sequence--a protein, in other words.[33]

Morris says of DNA: Higher organisms are composed of numerous specialized cells; and within each cell is an intricate complex of specialized protein molecules, each of which is a particularly organized structure with a composition of about twenty amino acids, each one made up of the four elements--hydrogen, oxygen, nitrogen, and carbon (in two cases sulfur is also present).[34]

The DNA molecule not only has information required for the synthesis of the specific protein molecules needed by the cell, but also information needed for its own replication. Therefore, reproduction and inheritance depend directly on this remarkable molecule, as organized differently and specifically for each kind of organism.[35]

Thus the problem of abiogenesis devolves upon the method by which the first replicating system evolved. The insuperable barrier, however, in that DNA can only be replicated with the specific help of certain protein molecules (enzymes) which in turn can only be produced at the direction of DNA. Each depends on the other and both must be present for replication to take place.

Thus, special creation seems to be the only way to account for the initiation of this process.[36]

Others believe that directions for the reproduction of plans, for energy and the extraction of parts from the current environment, for the growth sequence and for the effector mechanism translating instructions into growth--all had to be simultaneously present at that moment. This combination of events has seemed an incredibly unlikely happenstance, and has often been ascribed to divine intervention.

Still other writers have said that the most sweeping evolutionary questions at the level of biochemical genetics are still unanswered. The fact that in all organisms living today, the processes both of replication of the DNA and of the effective translation of its code require highly precise enzymes and that, at the same time, the molecular structures of those same enzymes are precisely specified by the DNA itself, poses a remarkable evolutionary mystery.

Did the code and means of translating it appear simultaneously in evolution? It seems almost incredible that any such coincidence could have occurred, given the extraordinary complexities of both sides and the requirement that they be co-ordinated accurately for survival. By a pre-Darwinian (or a skeptic of evolution after Darwin) this puzzle surely would have been interpreted as the most powerful sort of evidence for special creation.

Even the simplest of molecules is highly complex. Even if such a molecule could be formed by chance happening, it would be unable to reproduce itself. The fact that the DNA molecule is needed for reproduction and that it can only operate in the presence of proteins which are specified and organized, seems an impenetrable barrier to this vital phase of evolution.

Creationists have no problem with this because the creation model predicts life comes only from life.

The recapitulation theory applied to man for example taught that the human embryo began life as a marine protozoan, proceeded to a worm with a tube heart, to a fish, to an amphibian, etc., and passed through these stages of development to ultimately reach the human being.

The DNA for the man is not the same as the DNA for a fish, nor is it the DNA for a fish with something added. The DNA for each kind is programmed uniquely to produce its own kind and not to produce a temporary replica of some other kind.

The fact that the DNA molecule is basic in the reproductive mechanisms for all kinds of organisms is used to suggest common ancestry and offer evidence for evolution. The more significant fact is that each kind of organism has its own DNA molecular structure different from every other kind. Recalling the great complexity of DNA molecules, we must see that such a system could not have evolved by chance. One type of DNA cannot evolve into the DNA for another type of organism, its structure being designed to prohibit that very thing.

LIMITED VARIATION AND UNLIMITED CHANGE

Variation within limits in a certain kind of organism is clearly evident in domestic plants and animals. For example, in the species of Zea Mays, there are three kinds of corn man plants--sweet corn, popcorn, and field corn. They are all of the species Zea Mays and hybridize freely when grown near each other. There are red and white onions, yellow and white peaches and many kinds of roses. These exemplify plant variations within limits. There are many varieties within species of animals--many kinds of horses, dogs, and cows, which illustrate limited variation. Man can and has bred these variations by artificial selection to get the result he desires. In order to maintain desirable types of variations, selective breeding has to be controlled by man. These variations, however, do not demonstrate the origin of distinct new kinds or forms of life.

There are two ways that variation within limits are thought to occur. They are: by gene shift and natural selection.

Gene shift is the isolation of certain specific genes into one population and other genes into a different population. In dogs (all one species) the Chihuahua is very small and has short hair. Large breeds can either have long or short hair. By the use of artificial selection man can combine the genes desired in certain strains. If dogs were allowed to mate at will, the pure strains would disappear and the dogs could be highly variable mongrels. This fact applies to cows, roses, corn and other organisms.

In nature there are examples of natural selection resulting in shifts of genetic composition. One example is the peppered moth mentioned earlier. The change involved is an example of gene shift from mainly light to dark coloration. Another example is that light colored mice and lizards live abundantly on the white sands of New Mexico, and the darker animals of the species live on dark rocks nearby. It has been found by investigators that the processes of natural selection can convert an entire population into a new variety. Both artificial and natural selection shifts genes that are already present. There are new or different genes created by either one of these selection processes, which is necessary if unlimited change in organisms was to take place to the extent suggested by evolutionists. No matter what color the peppered moth is, it is still called Biston betularia. Similarly, domesticated plants and animals still belong to the species in spite of the great variation that occurs due to man's use of artificial selection. Selection has limitations and may cause changes beyond a certain point.

Polyploidy

In plants (not animals) polyploidy is thought by some to be a means of evolution. Polyploidy is defined as a condition where three or more monoploid sets of chromosomes are together in the cells of an organism. Usually in most organisms gametes are characterized by a single (monoploid) set of chromosomes. Usually, when two gametes unite to form a zygote, the new cell will then contain the diploid set of chromosomes, a set from each parent. Polyploidy is a condition whereby some plants will have variations of three or more homologous chromosome sets.

Polyploidy has many effects on plants. Polyploid cells are usually larger, which can result in a change of water content as one effect. The size of flowers and leaves can be affected by becoming larger.

Man can produce polyploids artificially by the use of colchicine, by the use of wounds, temperature shock, radiation, and sometimes hybridization. A famous polyploid produced by man is the cabbage radish hybrid. It was produced by Karpechenko, a Russion Biological researcher. Radishes and cabbages both have eighteen chromosomes in their body cells.

Karpechenko bred a hybrid containing thirty-six chromosomes--the tetraploid number. At the top of this strange hybrid was a radish, and the root was cabbage. Agriculturally it had no value. It was sterile and was unable to cross with parent radish or cabbage plants. It was considered as a new genus and was called Raphanobrassica which means radishcabbage. These plants soon died out, and Karpechenko's experiment has not been repeated by other scientists with the same results. A characteristic of a good scientific experiment is that it can be verified by other scientists.

There are several examples of multiple sets of chromosomes among both wild and cultivated plants. Some examples are wheat, some types of wild grasses, tobacco, and potatoes. There is no way to know how they occurred or whether such plants have always had the characteristic, except in those cases where man produced the polyploid. Attempts to explain these states can only be speculation.

Three important things about polyploids should be noted:

1. No mutation of a gene is involved; so no change of distinct kinds could result from this increase in number of already-present genes.

2. Artifically produced polyploids are less viable than the usual diploid forms and so produce fewer seeds.

3. It is known that doubling the chromosome number does not produce an entirely new organism. Neither does this production show how a single-celled algae could change into an oak tree.

Darwin's Finches

One famous example of variation which is considered as "evidence" of evolution by proponents is the Darwin's finches that Darwin found on the Galapagos Islands while on his voyage on the H.M.S. Beagle. Darwin believed that these finches on a small scale showed how evolution has occurred. He proposed that all the finches he placed in at least four genera came from a common ancestral pair of birds (adaptive radiation). He suggests that the ancestors had flown to the Galapagos Islands from the South American mainland. The chief differences in the Darwin Finches are beak and body size. The females are a drab gray-brown and the males are mostly black.

Darwin felt that the birds that had larger beak size would be more equipped to eat large seeds. The larger beaks could more easily crush seeds to make them more digestible. He believed those with smaller beaks adapted to an insect diet. The long-beaked birds gradually became adapted to eating undeveloped seeds from inside the cactus fruit.

Recently more than 1,200 specimens of Darwin's finches have been restudied at the California Academy of Science Museum. Perhaps Darwin was enthusiastic in regard to the amount of variation involved. Even though they are still classified into the genera Geospiza, Platyspiza, and Camarhyncous, all finches of these genera and the assigned species integrate with each other. In other words, if the labels of the collection were removed and the birds arranged according to body size and beak size, there would be found a perfect graduation between the largest-beaked species Geospiza magnirostris and the species with the smallest beak, G. fuliginosa. It does seem though that most specimens of G. scandens, that have quite a long beak, are truly distinctive (they eat cactus fruit); yet some individuals of G. scandens are identical with the ones with the longest beak in the G. conirostris variety which integrates with G. Magirostris.

Darwin's finches are a highly variable species that probably populated the islands originally as a flock of finches. Their chance settlement of the various islands set up a distribution where only G. scandens thrive on certain islands, and G. magnirostris on the other islands and other genera in similar random distribution. The basic song pattern is reported to be the same with little variation in the different genera.

So far, attempts to hybridize various groups of finches have not been successful, though it is true that animals often do not reproduce in captivity. The degree of intergradation is remarkable; and, if it were not for the historical value placed on these birds (because of Darwin's association with them), it is doubtful that they would still be retained as true species.

Instead of being a problem, these birds are an example of genetic principles. If some animals are isolated, latent genes will become expressed and different or various characteristics develop. This occurs because animals mate with others also possessing the latent genes. If the finches remained on the South America mainland, many of the new characteristics of color and beak shape may not have been fully expressed. They would have continued to mate with birds having dominant genes and this would have hidden those characteristics. All types of human beings are assigned to one species, yet they demonstrate wider variation than Darwin's finches and are classified as several species. These contrasting situations provide ample evidence that the classification schemes prepared and used by scientists are somewhat arbitrary.

Natural Selection and Complexity

Natural selection and complexity is a subject of importance. It is believed by evolutionists that the remarkable complexity found in living things has been built up by natural selection processes. We know that no two living things are alike exactly. But if the genes are alike, as in identical twins, or in a pure line of beans, the differences in the individuals are due to environment. The differences are being passed on to the next generation, and so do have genetic significance.

When Darwin formulated the theory of evolution by natural selection, a hundred years ago, he knew nothing about genes or even their environmental changes (acquired characteristics) which are not inherited. We know now that no organism is simple, even though it was once postulated that life began from simple plants and animals. Some of them by chance became more complex and had an advantage over others, and the simple plants and animals became extinct. This, in essence, is how evolutionists use natural selection to account for the development of complex organisms.

In many instances, it can be seen that less complex organisms survive over more complex ones. We might wonder how well animals in their natural environment fit the natural selection theory to explain development of complexity. In one instance, the lesser complex hydra is predator to and feeds on the more complex Daphnia which has legs, eyes, brain, antenae, heart, alimentary canal, and an abdominal claw, in comparison to hydras two-layer sac. The less complex selects the higher complex form. Another example, though, in the plant kingdom is the aster family, the daisies which have the most complex structure of any plant. They are crowded out by jack pines just because the latter are greater in size.

The creationists suggest that though we do not know what the exact appearance of living things was at the time of creation, it seems reasonable to estimate that living things were created with much more potential variability. They had many latent genes which became expressed later making up the groups called (by different persons) varieties, races or species. The explanation of the finch variation has wide application.

The evolutionist asks the question: "How do you know this?"

FOOTNOTES

[32]Katherine P. Anthony, Textbook of Anatomy and Physiology, 8th Ed. (Missouri: C.V. Mosby Company, 1971), p. 426

[33]Ibid.

[34]Henry M. Morris ed., Scientific Creationism, (California: Creation Life Publishers, 1974), p. 47.

[35]Ibid.

[36]Ibid.

CHAPTER V

CONCLUSION

They called John T. Scopes an infidel. He was accused of teaching school children that man likely evolved from other forms of life. Creationists found this contradictory to the story of creation as revealed in the Book of Genesis.

While Scopes' conviction was overturned by the Tenessee Supreme Court on a technicality, he did not consider the argument won. "The fight will still go on with other actors and other plays," he later said.

Long after the monkey trial drama that lifted him to Bible Belt infamy, Scopes' prediction proved true. The curtain rose in December, 1981, on another act in the long battle between creationists and evolutionists over what children should learn about the beginning of mankind: an American Civil Liberties Union suit attacked a creationist-backed Arkansas law and won. . . for <u>now</u>.

But Scopes would have difficulty recognizing the battlelines of today. Creationists are now espousing one of the arguments of Clarence Darrow's ardent defense of Scopes: that the theory of the beginning should not be taught to the exclusion of another. In short, perhaps creationism should be considered along with the evolutionary theory.

With that thought in mind, the author has written this book giving arguments for and against the evolution theory. An attempt has been made to espouse the beliefs of the creationists using scriptures in their defense, but keeping in mind that there is no <u>scientific</u> data that supports creationism, nor is there a professional organization of scientists in America, other than the Creation Research Society, that believes there is any scientific support for creationism.

Thus, this book promises to fill a great and growing need for a new dimension in this controversy: an opportunity for our students to analyze the controversy critically and to reach their own conclusions.

BIBLIOGRAPHY

Alexander, R. D. "Evolution, Creation, and Biology Teaching." American Biology Teacher, 40 (2): 91-104, 1978.

Anthony, Katherine P. Textbook of Anatomy and Physiology. 8th ed. Missouri: C. V. Mosby Company, 1971.

Barnes, Thomas G., and Davidheiser, Bolton. And God Created (series)--Origins Without God, The Beginning of Life, and Theistic Evolution. California: Creation Life Publishers, 1973.

Bolton, Davidheiser. And God Created--Theistic Evolution. California: Creation Life Publishers, 1973.

Brash, S. G. "Creationism/Evolution: The Case Against Equal Time." The Science Teacher, 48 (4): 29, 1981.

Calagan, C. A. "Evolution and Creationists' Arguments." American Biology Teacher, 42 (7): 422-425, 1980.

Chittick, Donald E. Creation Counseling--The Age of the Earth. California: Creation Life Publishers, 1973.

Cory, L. R. "Creationism and The Science Method." American Biology Teacher, 35 (4): 223-225, 1980.

Dean, H. Douglass. Creation Counseling--Life in a Test Tube. California: Creation Life Publishers, 1973.

Dobzhansky, T. Genetics and "The Origin of Species." 3rd ed. New York: Columbia University Press, 1951.

Dobzhansky, T. "Nothing in Biology Makes Sense Except in The Light of Evolution." American Biology Teacher, 35 (3): 125-129, 1973.

Howe, Dr. George. And God Created--Louis Pasteur. California: Creation Life Publishers, 1973.

Holtzman, Eric, and Novikoff, Alex B. Cells and Organelles. Holt, Rinehart, and Winston, Inc., 1970.

Hughes, Stuart W. "The Fact and The Theory of Evolution." American Biology Teacher, 44 (1): 25-31, 1982.

Hyma, Albert and Stanton, Mary. Streams of Civilization. California: Creation Life Publishers, 1976.

Levine, Louis. Biology of the Gene. Missouri: C. V. Mosby Company, 1969.

Miller, Kenneth R. "Special Creation and The Fossil Record: The Central Fallacy." American Biology Teacher, 44 (2): 85-89, 1982.

Milne, D. H. "How to Debate with Creationists--and Win." American Biology Teacher, 35 (3): 235-245, 1981.

Moore, John N. and Slusher, Harold S., ed. Biology --A Search for Order in Complexity. Michigan: Zondervan Corporation, 1976.

Moore, John N. And God Created--Impact of Evolution on Society and Evolution and Communism. California: Creation Life Publishers, 1973.

Morris, Henry M. Biblical Cosmology and Modern Science. New Jersey: Craig Press, 1975.

Morris, Henry M. ed. Scientific Creationism. California: Creation Life Publishers, 1974.

Morris, Henry M. The Remarkable Birth of Planet Earth. California: Institute for Creation Research, 1973.

Rosenfeld, R. P. "Antievolutionary Misconceptions." American Biology Teacher, 39 (a): 547-549, 1977.

Sherman, I. and Sherman, V. Biology: A Human Approach. New York: Oxford University Press, 1975.

Skoog, A. "Does Creationism Belong in The Biology Curriculum?" American Biology Teacher, 40 (1): 23-26, 1978.

Stephens, G. and North, B. Biology. New York: J. P. Wiley and Sons, 1974.

Whitcomb, John C. And God Created--Creation of the World and Methods of the Creator. California: Creation Life Publishers, 1973.

Whitcomb, John C. And God Created--Days of Creation and The Gap Theory. California: Creation Life Publishers, 1973.

Whitcomb, John C. Creation Counseling--The Origin of Man. California: Creation Life Publishers, 1973.

Whitcomb, John C., Jr. The Early Earth. Michigan: Baker Book House, 1977.

GLOSSARY

Adaptation: Evolving characteristics that make a population or individual better suited to its environment.

Analogous: A term that describes structures that serve the same function but have different morphologies and embryonic origins.

Autotroph: Self feeder, able to manufacture food such as through photosynthesis.

Biogenesis: The theory that living things can be produced only by other living things.

Catastrophism: The theory that certain geological and biological phenomena were caused by catastrophes rather than by continuous and uniform processes.

Cell: The basic unit of life. The Simplest unit that can exist as an independent system.

Chromosome: Thread-like, condensed chromatin (DNA), visible at cell division only. Occurring in pairs with specific numbers for each species.

Convergent Evolution: A similarity in genetically different organisms resulting from their adaptation to similar habitats.

Deoxyribonucleic Acid: DNA, a nucleic acid containing sugar deoxyribose, which is found chiefly in the nucleus of cells. DNA is the genetic message, functioning in protein systhesis.

Dominant: In genetics, the allele of a pair that is expressed while the other is hidden or repressed.

Embryology: The branch of biology that deals with the formation and development of embryos.

Evolution: The continuous genetic changes occurring in populations as a result of adapting to environmental changes.

Extinct: No longer in existence.

<u>Fossil</u>: The hardened remains or traces of an animal or plant of a former age.

<u>Heredity</u>: The transmission of traits from parents to offspring.

<u>Invertebrates</u>: Any of the large number of types of animals without a backbone.

<u>Mechanistic Theory</u>: The theory that the processes of life are based on the same physical and chemical laws that apply to the non-living phenomena.

<u>Mutation</u>: Any change in a gene that results in a permanent genetic change in the individual; a deletion or addition to the basic code of the DNA strand.

<u>Natural Selection</u>: The differential survival and reproduction of certain individuals in a population. In the same process, other organisms less suited to the environment are eliminated.

<u>Phylum</u>: The major primary subdivision of a kingdom. It is composed of classes (exception in plant groups).

<u>Polyploid</u>: Having a chromosome number that is three or more times the haploid number.

<u>Ribonucleic Acid (RNA)</u>: A nucleic acid containing ribose sugar and the nitrogen base, uracil. Functions in protein synthesis.

<u>Scientific Method</u>: An orderly method used by scientists to solve problems in which a recognized problem is subjected to thorough investigation, and the resulting facts and observations are analyzed, formulated in a hypothesis, and subjected to verification by means of experiments and further observations.

<u>Sex-linked</u>: A trait determined by a gene located on a sex chromosome.

<u>Species</u>: The major subdivision of a genus. Individuals of a species are able to breed among themselves under natural conditions.

<u>Spontaneous Generation</u>: The notion that living organisms are derived from non-living matter.

<u>Taxonomy</u>: The science of classifying life.

Uniformitarianism: The theory that geological change is caused by a gradual process rather than by a sudden upheaval.

Vertebrates: Belonging to, or having to do with a group of animals that have a backbone, or segmented spinal column, and a brain case, or cranium enclosing the brain.

Vitalism: The notion that life has unique mystical properties that are distinct from those ascribed by chemical and physical laws.

Biographical Data

Adell Thompson, Jr. Associate Professor of Biology and Science Education teaches at the University of Missouri-Kansas City. He has taught for the last thirteen years at the university level. Dr. Thompson has taught General Biology, Anatomy, Physiology, Lab Techniques in Biology Instruction, Secondary Methods Science and supervised student teachers in Science and Math. At the present time, Dr. Thompson is State Director of the Missouri Academy of Science, Director of the Student Teaching Program and Coordinator of Educational Placement at the University of Missouri-Kansas City and teaches a course in Life Science. Before coming to the University of Missouri-Kansas City, Dr. Thompson taught Chemistry, Biology, Human Science Physiology, and General Science at the high school level for eleven years. He was also the chairman of the science department for three years and spent his last year in the public school system as a science consultant. He has written numerous articles in major science journals on science curriculum and instruction.